MASTERING
ELECTRICAL ENGINEERING

NOEL M. MORRIS

MACMILLAN

First published 1985

Published by
MACMILLAN EDUCATION LTD
Houndmills, Basingstoke, Hampshire RG21 2XS
and London
Companies and representatives
throughout the world

Printed and bound in Great Britain by
Anchor Brendon Ltd, Tiptree, Essex

British Library Cataloguing in Publication Data
Morris, Noel M.
Mastering electrical engineering.—(Macmillan
master series)
1. Electric engineering
I. Title
621.3   TK145
ISBN 0-333-38592-6
ISBN 0-333-38593-4 Pbk
ISBN 0-333-38594-2 Pbk (export)

CONTENTS

CONTENTS

CONTENTS

CONTENTS

CONTENTS

LIST OF TABLES

LIST OF FIGURES

PREFACE

In common with other books in the *Mastering* series, the intention is to set out in one book the basis of a complete subject – *Electrical Engineering* in this volume.

This book reflects the modern thinking and needs of the *electrical technologist*, emphasis being placed on practical circuits and systems. Following modern teaching practice in many courses, the mathematics has been kept to a minimum consistent with covering the necessary background theory.

Mastering Electrical Engineering is suitable for use as a self-teaching book, each chapter being supported not only by worked examples but also by self-test questions and a summary of important facts. The latter feature is very useful for the reader who is in a hurry to get a 'feel' for the subject matter in the chapter.

Starting with the principles of electricity and sources of *electromotive force* (e.m.f.), the book covers the basis of 'heavy current' electrical engineering including *circuit theory*, *alternating current* (a.c.) and *direct current* (d.c.) *machines*, *single-phase* and *three-phase* calculations, *transformers*, *electrical power distribution*, *instruments* and *power electronics*.

The book contains a liberal supply of illustrations to highlight the features of each chapter, and it is hoped that the approach will stimulate the reader with the same enthusiasm for the subject that the author holds for it.

I would like to thank the electrical manufacturing industry at large for the support it has given, particular thanks being due to the following:

> RS Components (for figs 16.2, 16.3 and 16.10c)
> GEC Machines
> Celdis Ltd.
> AVO Ltd

The author and publishers wish to thank the Southern Electricity Board for permission to use the cover illustration of one of the Board's sub-stations.

The author would also like to thank Elizabeth Black and Keith Povey, through Macmillan Education, for their invaluable assistance in tidying up the manuscript.

Finally, I must thank my wife for her help, patience and support during the months whilst the book was being written.

North Staffordshire Polytechnic   NOEL M. MORRIS

DEFINITIONS OF SYMBOLS USED IN EQUATIONS

A, a	area in m^2
B	magnetic flux density in tesla (T)
C	capacitance in farads (F)
C	unit of electrical charge (coulombs)
D	electric flux density in coulombs per square meter (C/m^2)
d	diameter and distance in metres (m)
E, e	e.m.f. or p.d. in volts (V)
E	electric field intensity or potential gradient in volts per metre (V/m)
e	base of Naperian logarithms = 2.71828
F	magnetomotive force in ampere-turns or in amperes (A)
F	mechanical force in newtons (N)
F	unit of capacitance (farads)
f	frequency in hertz (Hz)
f$_o$	resonant frequency in hertz (Hz)
G	conductance in siemens (S)
H	magnetic field intensity or magnetising force in ampere-turns per metre or amperes per metre (A/m)
H	unit of inductance (henrys)
I, i	current in amperes (A)
K	a constant of an electrical machine
k	magnetic circuit coupling coefficient (dimensionless)
L	self-inductance of a magnetic circuit in henrys (H)
l	length in metres (m)
M	mutual inductance between magnetic circuits in henrys (H)
N	number of turns on a coil
N, n	speed of rotation of the rotating part of a motor in revolutions per minutes (rev/min) or revolutions per second (rev/s)
P	power in watts (W)
Q	electric charge of electrostatic flux in coulombs (C)

Q	Q-factor of a resonant circuit (dimensionless)
Q	reactive volt-amperes (VAr) in an a.c. circuit
R, r	resistance in ohms (Ω)
S	magnetic circuit reluctance (resistance to flux) in ampere-turns per weber or amperes per weber (A/Wb)
S	shunt resistance connected to a meter
S	volt-amperes (VA) in an a.c. circuit
s	fractional slip of an induction motor rotor (dimensionless)
T	periodic time of an alternating wave in seconds (s)
T	time constant of an electrical circuit in seconds (s)
T	torque in newton metres (N m)
t	time in seconds (s)
V, v	voltage or p.d. in volts (V)
W	energy in joules (J) or in watt seconds (W s)
X_C	capacitive reactance in ohms (Ω)
X_L	inductive reactance in ohms (Ω)
Z	impedance of an a.c. circuit in ohms (Ω)
α	temperature coefficient of resistance in $(^{\circ}C)^{-1}$
ϵ	absolute permittivity of a dielectric in farads per metre (F/m)
ϵ_O	permittivity of free space $= 8.85 \times 10^{-12}$ F/m
ϵ_r	relative permittivity of a dielectric (dimensionless)
η	efficiency of an electrical machine
θ	temperature in $^{\circ}$C or K
θ	angular measurement in degrees or radians
μ	absolute permeability of a magnetic material in henrys per metre (H/m)
μ_O	permeability of free space $= 4\pi \times 10^{-7}$ H/m
μ_r	relative permeability (dimensionless)
π	a constant $= 3.142$
ρ	resistivity of an electrical conductor in ohm metres (Ω m)
σ	conductivity of a conductor in siemens per metre (S/m)
Φ	magnetic flux in webers (Wb)
ϕ	phase angle in degrees or radians
$\cos \phi$	power factor of an a.c. circuit
ω	angular frequency in rad/s of an a.c. supply
ω	speed of rotation of the rotating part of an electric machine in rad/s
ω_O	resonant frequency in rad/s

GLOSSARY

Words in *italics* are mentioned elsewhere in the Glossary.

a.c. An abbreviation for *alternating current*

acceptor circuit A *series resonant circuit* which has a very low resistance to current flow at the *resonant frequency*, that is, it 'accepts' a high current

a.c. machine An electromechanical energy convertor which converts energy from an *a.c.* source into mechanical energy or vice versa

accumulator An electrical *storage battery*, that is, a battery which can be recharged by passing *direct current* through it

alternating current A current which alternately flows in one direction and then in the opposite direction

alternator An alternating current *generator*

ammeter An instrument for the measurement of electrical *current*

ampere The unit of electrical *current*

ampere-turn The unit of *magnetic field intensity* (H) or *magnetising force*, which is calculated from amperes × turns on the coil; since 'turns' are dimensionless, it is given the unit of the 'ampere' by electrical engineers

anode (1) In a *diode* it is the *electrode* by which the *current* (*hole* flow) enters; (2) In *electrolysis*, it is the electrode to which negative *ions* are attracted

apparent power In an a.c. *circuit*, it is the product, volts × amperes (or the volt-ampere [VA] product)

armature (1) The rotating part of a *d.c. machine*; (2) In a relay, it is a piece of ferromagnetic material which is attracted towards the pole of the electromagnet

autotransformer A *transformer* having a single winding

average value The average value of an alternating wave. An alternative name is *mean value*

back e.m.f. The e.m.f. induced in an *inductor* when the *current* through it changes

battery A group of *cells* connected together

brush A piece of specially shaped carbon or graphite which connects either the *commutator* of a *d.c. machine* or the *rotor* of an *a.c. machine* to the external circuit

cage rotor motor A popular form of *induction motor* in which the *rotor* consists of metal rods (copper or aluminium) embedded in a *laminated* iron circuit, the bars being short-circuited by means of 'end rings' at the ends of the rotor

capacitance The property of a *capacitor* which enables it to store electrical charge

capacitive reactance The opposition of a *capacitor* to the flow of *alternating current*. No *power* is dissipated in a pure capacitive reactance. Symbol X_C, measured in ohms

capacitor Consists of two conducting surfaces or 'plates' separated by an insulating *dielectric*, which has the ability to store electric charge

cathode (1) In a *diode*, it is the *electrode* by which the *current* (*hole* flow) leaves; (2) In *electrolysis*, it is the electrode to which the positive *ions* are attracted

cell Converts chemical energy into electrical energy

circuit An interconnected set of *conductors*

coercive force The *magnetising force* needed to *demagnetise* completely a piece of magnetised material

commutator Consists of a large number of conducting segments connected to the *armature* winding of a *d.c. machine*, each segment being isolated from adjacent segments; Current enters the armature via graphite *brushes*

complex wave A wave which contains a *fundamental frequency* together with a number of *harmonic frequencies*

compound-wound machine A *d.c. machine* having part of its *field winding* in *series* with its *armature*, and part connected in *shunt* with the armature

conductance Reciprocal of *resistance*. Symbol G, and measured in siemens (S)

conductivity Reciprocal of *resistivity*

conductor An element which freely allows the flow of electric *current*

core loss Energy loss in an electrical machine as a result of the combined effects of *hysteresis loss* and *eddy current* loss

coulomb The unit of electrical charge, symbol C

current Rate of flow of electrical charge. Symbol I, and measured in *amperes* (A)

d.c. Abbreviation of *direct current*

d.c. machine An electromechanical energy convertor which converts energy from a *d.c.* source into mechanical energy or vice versa

depolarising agent A chemical included in a *cell* to prevent *polarisation*

dielectric An insulating material which separates the plates of a *capacitor*

diode A two-electrode device, the electrodes being the *anode* and the *cathode*

direct current *Current* which flows in one direction only, that is, a unidirectional current

eddy current *Current* induced in the iron circuit of an electrical machine because of changes in *magnetic flux*

efficiency Ratio of the power output from a machine or circuit to its input power; expressed as a per centage if the ratio is multiplied by 100, and is dimensionless

electric field intensity The potential gradient in volts per metre in the material

electric flux A measure of the electrostatic field between two charged plates; measured in coulombs

electric flux density The amount of *electric flux* passing through one square metre of material

electrode (1) In a *semiconductor* device it is an element which either emits *current* or collects it; (2) In an electrolytic cell it is a metallic conductor where the *ions* give up their charge

electrolysis A chemical change brought about by the passage of *direct current* through an *electrolyte*

electrolyte A substance which, when dissolved, produces a conducting path in the solvent (which may be water)

electromagnet A current-carrying coil with an iron core

electromagnetic induction The production of an *e.m.f.* in a circuit, arising from a change in the amount of *magnetic flux* linking the circuit

electromotive force The *p.d.* measured at the terminals of a *cell, battery* or *generator* when no *current* is drawn from it; abbreviated to *e.m.f.* and measured in *volts*

electron A negative charge carrier, and a constituent part of every atom

e.m.f. Abbreviation for *electromotive force*

energy meter A meter used to measure energy, usually in kilowatt hours (kWh)

exciter A d.c. *generator* which provides the *current* for (that is, it 'excites') the *field winding* of an *alternator* or *synchronous motor*

farad The unit of *capacitance*, symbol F; submultiples such as the microfarad, the nanofarad and the picofarad are in common use

Faraday's laws (1) The laws of *electrolysis* relate to the mass of substance liberated in the process of electrolysis; (2) the law of *electromagnetic induction* relates to the induced *e.m.f.* in a circuit when the *magnetic flux* associated with the circuit changes

ferromagnetic material A material which can be strongly magnetised in the direction of an applied *magnetising force*

field winding A winding on an electrical machine which produces the main magnetic field

Fleming's rules The left-hand rule relates to *motor* action, the right-hand rule relates to *generator* action

frequency The number of oscillations per second of an alternating wave;

measured in hertz (Hz)

full-wave rectifier A circuit which converts both the positive and negative half-cycle of an *alternating current* wave into *direct current* (more precisely, unidirectional current)

fundamental frequency The *frequency* of a sinusoidal wave which is the same as that of the complex wave of which it is a part

galvanometer A moving-coil meter used to measure small values of current

generator An electromechanical energy convertor which changes mechanical energy into electrical energy

half-wave rectifier Converts one of the half-cycle of an *a.c.* waveform into *direct (unidirectional) current*, but prevents current flow in the other half cycle

hard magnetic material A material which retains much of its magnetism after the *magnetising force* has been removed

harmonic frequency A multiple of the *fundamental frequency* of a *complex wave*

henry Unit of inductance, symbol H

hertz Unit of *frequency*, symbol Hz; equal to 1 cycle per second

hole A positive charge carrier; can be regarded as the absence of an *electron* where on would normally be found

hysteresis loss Energy loss caused by the repeated reversals of magnetic domains in a *ferromagnetic material* in the presence of an alternating *magnetic field*

impedance Total opposition of a *circuit* to the flow of *alternating current*; symbol Z, measured in *ohms*

induced e.m.f. *e.m.f.* induced in a *circuit* either by a changing *magnetic flux* or by a strong electric field

inductance A measure of the ability of a *circuit* to produce a *magnetic field* and store magnetic energy

induction motor An a.c. motor which depends for its operation on a 'rotating' or 'moving' *magnetic field*

inductive reactance The opposition of a pure *inductance* to the flow of *alterating current*; no *power* is dissipated in an inductive reactance; symbol X_L, measured in *ohms*

inductor A piece of apparatus having the property of *inductance*

instrument transformer A *transformer* designed to connect an electrical instrument either to a high *voltage* (a voltage transformer, VT, or potential transformer, PT) or to a high *current* (a current transformer, CT)

insulator A material which has a very high *resistance* to the flow of electrical *current*. Ideally, no current flows through an insulator

internal resistance The *resistance* 'within' a *cell, battery, generator* or power supply

invertor A circuit which converts direct voltage or *direct current* into alternating voltage or *alternating current*

ion An atom or molecule which is electrically charged; can be either negatively or positively charged

ionisation The process by which an atom or molecule is converted into an *ion*

joule The unit of energy equal to 1 watt × 1 second

junction The connection of two or more wires in a circuit; *node* is an alternative name

Kirchhoff's laws (1) The total *current* flowing towards a *junction* is equal to the total current flowing away from it; (2) the algebraic sum of the *p.d.s.* and *e.m.fs* around any closed mesh is zero

lamination A thin sheet of iron, many of which are grouped together to form a **magnetic circuit**; used to reduce *eddy current*

magnetic circuit An interconnected set of ferromagnetic branches in which a *magnetic flux* is established

magnetic coupling coefficient A dimensionless number having a value between zero and unity which gives the proportion of the *magnetic flux* which arrives at a second (*secondary*) coil after leaving the *primary* winding; symbol k

magnetic domain A group of atoms in a *ferromagnetic material* which form a localised magnetic field system

magnetic field intensity The *m.m.f.* per unit length of a *magnetic circuit*; symbol H; measured in ampere-turns per metre or amperes per metre

magnetic flux A measure of the magnetic field produced by a *permanent magnet* or *electromagnet*; symbol Φ; measured in webers (Wb)

magnetic flux density The amount of *magnetic flux* passing through an area of 1 m^2; symbol B, measured in tesla (T)

magnetic leakage *Magnetic flux* which does not follow the 'useful' magnetic path

magnetic leakage coefficient The ratio of the total *magnetic flux* to the 'useful' magnetic flux; has a value 1.0 or greater

magnetising force An alternative name for *magnetic field intensity*

magnetomotive force The 'force' which produces a *magnetic flux*; symbol F, measured in ampere-turns or amperes; abbreviation m.m.f.

mean value The *average value* of an alternating wave

motor An electromechanical energy convertor which changes electrical energy into mechanical energy

mutual inductance The property of a system which causes a change of *current* in one circuit to induce a *voltage* in another circuit

negative charge carrier An *electron*

node Alternative name for *junction*

non-linear resistor A *resistor* which does not obey *Ohm's law*

n-type semiconductor A *semiconductor* material which has been 'doped' so that it has mobile *negative charge carriers*

ohm The unit of electrical *resistance* or *impedance*, symbol Ω

ohmmeter A moving-coil instrument used to measure *resistance*

Ohm's law This states that, at a constant temperature, the *current* in a pure *resistor* is directly proportional to the *p.d.* across it

parallel circuit A circuit in which all the elements have the same *voltage* across them

parallel resonant circuit An a.c. *parallel circuit* containing *resistance*, *inductance* and *capacitance* which resonates with the supply frequency; known as a *rejector circuit*, it has a high *impedance* at *resonance*, and the circulating *current* within the *circuit* is higher than the supply current

p.d. Abbreviation for *potential difference*

Peltier effect When a *current* flows in a *circuit* consisting of dissimilar *semiconductors* or metals, the Peltier effect describes why one junction is heated and the other is cooled

periodic time The time taken for one cycle of an *a.c.* wave to be completed

permanent magnet A piece of *ferromagnetic material* which has been permanently magnetised. Both its *remanence* or *retentivity* and its *coercive force* are high

permeability The ratio of the *magnetic flux density* (B) in a material to the *magnetic field intensity* (H) needed to produce it. Also known as the absolute permeability of the material. Symbol μ, measured in henrys per metre (H/m)

permeability of free space The permeability of a vacuum (or, approximately, of air), symbol $\mu_O = 4\pi \times 10^{-7}$ H/m

permeability (relative) The ratio of the absolute *permeability* of a magnetic material to the *permeability of free space*; symbol μ_r, and is dimensionless

phase angle The angular difference in degrees or radians between two sinusoidally varying quantities or between two *phasors*

phasor A line which is scaled to represent the *r.m.s.* value of a waveform and whose angle represents its displacement from a phasor in the horizontal 'reference' direction

piezoelectric effect The production of an *e.m.f.* between two faces of a crystal when it is subject to mechanical pressure. The effect is reversible

polarisation A chemical effect in a *cell* which causes the *anode* to be coated with hydrogen bubbles

pole (1) A terminal of a *cell*; (2) one end of a *permanent magnet* or an *electromagnet*

poly-phase supply An *a.c.* supply having many ('poly') phases; the *three-phase supply* is the most popular type

positive charge carrier A *hole*

potential A measure of the ability of a unit charge to produce *current*

potential difference The difference in electrical *potential* between two points in a *circuit*

potentiometer (1) A *resistor* having a sliding contact; (2) a device or *circuit* for comparing electrical *potentials*

power The useful output from an electrical machine and the rate of doing work; symbol P, measured in *watts* (W) or joules per second

power factor The ratio in an *a.c.* circuit of the *power* in *watts* to the *apparent power* in *volt-amperes*

primary cell A *cell* which cannot be recharged

primary winding The winding of a *transformer* which is connected to the *a.c.* supply source

***p*-type semiconductor** A *semiconductor* material which has been 'doped' so that it has mobile *positive charge carriers (holes)*

Q-factor The 'quality' factor of a *resonant circuit*; it indicates, in a *series resonant circuit*, the value of the *voltage* 'magnification' factor and, in a *parallel resonant circuit*, the value of the *current* 'magnification' factor

radian An angular measure given by the ratio of the arc length to the radius; there are 2π radians in a circle

reactance The property of a reactive element, that is, a pure *capacitor* or a pure *inductor*, to oppose the flow of *alternating current; power* is not consumed by a reactive element

reactive volt-ampere Also known as reactive 'power'; associated with *current* flow in a *reactive* element; 'real' power is not absorbed; symbol Q, measured in volt-amperes reactive (VAr)

rectifier A circuit which converts alternating voltage or current into direct (unidirectional) voltage or current

rejector circuit A *parallel resonant circuit* which has a very high *resistance* to *current* flow at the *resonant frequency*, that is, it 'rejects' current

reluctance The ratio of the *magnetomotive force* (*F*) in a *magnetic circuit* to the *magnetic flux* (Φ) in the circuit; it is the effective resistance of the circuit to magnetic flux; symbol S, measured in ampere-turns per weber or in amperes per weber

remanence The remaining *magnetic flux* in a specimen of magnetic material after the *magnetising force* has been removed; also known as the *residual magnetism* or *retentivity*

residual magnetism Another name for *remanence*

resistance A measure of the ability of a material to oppose the flow of *current* through it; symbol R, measured in ohms

resistivity The *resistance* of a unit cube of material, calculated by resistance $\times \frac{\text{area}}{\text{length}}$. Symbol ρ, measured in ohm meters (Ω m)

resistor A circuit element having the property of *resistance*

resonance The condition of an *a.c.* circuit when it 'resounds' or resonates in sympathy with the supply *frequency*; the *impedance* of the circuit at this frequency is purely *resistive*

resonant frequency The *frequency* at which the circuit resonates. Symbol ω_O (rad/s) or f_O (Hz)

retentivity Another name for *remanence*

rheostat A variable *resistor*

r.m.s. Abbreviation for *root-mean-square*

root-mean-square The *a.c.* value which has the same heating effect as the equivalent *d.c.* value; abbreviated to *r.m.s.* and also known as the 'effective value'

rotor The rotating part of a machine (usually associated with *a.c. machines*)

saturation (magnetic) The state of a *ferromagnetic material* when all its *domains* are aligned in one direction

secondary cell A *cell* which can be recharged by passing *d.c.* through it

secondary winding The winding of a *transformer* which is connected to the electrical load

Seebeck effect The *e.m.f.* between two dissimilar metals when their junctions are at different temperatures

self-inductance An alternative name for *inductance*

semiconductor A material whose *conductivity* is mid-way between that of a good *conductor* and that of a good *insulator*

semiconductor junction A junction between an *n-type semiconductor* and a *p-type semiconductor*; a *diode* has one *p–n* junction, and a junction *transistor* has two *p–n* junctions

separately excited machine A *d.c.* machine whose *field windings* and *armature* are supplied from separate power supplies

series circuit A circuit in which all the elements carry the same current

series motor A *d.c. motor* whose *field windings* are connected in *series* with the *armature*; mainly used for traction applications

series resonant circuit An *a.c. series circuit* containing *reactive* elements which *resonates* with the supply *frequency*. Known as an *acceptor circuit*, it has a low *impedance* at resonance, so that the resonant current is high, i.e., it 'accepts' *current*. The *voltage* across each of the *reactive* elements (*inductance* and *capacitance*) is higher than the supply *voltage*

shunt An alternative name for *parallel connection*

shunt wound machine A *d.c. machine* whose *field windings* are connected in *shunt* (parallel) with the *armature*

single-phase supply An *a.c.* supply system carried between two lines, one line usually being 'live' and the other being 'neutral', that is, at 'earth potential'

slip (fractional) The difference between the synchronous speed of the rotating field and the rotor speed of an induction motor expressed as a ratio of the synchronous speed, symbol s, and is dimensionless

slip ring A metal ring which is on, but is insulated from, the shaft of a rotating electrical machine; its function is to convey *current* to or from the rotating part of the machine (usually an *a.c. machine*) via carbon *brushes*. An a.c. machine may have two or more slip rings

smoothing circuit A *circuit* which 'smooths' or 'filters' the variations in the output *voltage* from a rectifier circuit

soft magnetic material A *magnetic material* which easily loses its magnetism; its *coercive force* is low

solenoid A coil with an air core

squirrel cage motor Alternative name for a *cage rotor motor*

storage battery Alternative name for *accumulator*

synchronous motor An *a.c. motor* whose rotor runs at synchronous speed

terminal voltage The *p.d.* between the terminals of a *cell, battery* or *generator* when an electrical load is connected to it

tesla The unit of *magnetic flux density*, symbol T

thermistor A *semiconductor* device whose *resistance* changes with temperature (usually a decrease in resistance with increase in temperature, but could be an increase in resistance)

thermocouple A *junction* of two dissimilar metals which develops an *e.m.f.* when it is heated or cooled relative to the remainder of the circuit

thermopile Several *thermocouples* connected in *series* to give a higher *e.m.f.* than a single thermocouple

three-phase supply A *poly-phase supply* having three phases; a 'balanced' or 'symmetrical' three-phase supply has three equal voltages which are displaced from one another by $120°$

tesla The unit of *magnetic flux density*, symbol T

thyristor A four-layer *semiconductor* device for the control of 'heavy' *current* by electronic means

torque Turning moment (force × radius) produced by a rotating machine

transformer Device which 'transforms' or changes *voltage* and *current* levels in an *a.c.* circuit; uses the principle of *mutual inductance*

transient A phenomenon which persists for a short period of time after a change has occurred in the circuit

transistor A three-layer *semiconductor* *n–p–n* or *p–n–p* device used in electronic amplifiers and computers

triac A multi-layer *semiconductor* device for the bidirectional control of *current* by electronic means; it is one of the *thyristor* family of elements

two-part tariff   A method of charging for the cost of electricity; one part of the tariff is related to the 'standing' charges associated with the generating plant, the other being related to the 'running' charges

universal motor   A *motor* which can operate on either *a.c.* or *d.c.*; a hand-held electrical drilling machine is an example

volt   The unit of *electromotive force* and *potential difference*, symbol V

voltage dependent resistor   A *resistor* whose *resistance* is dependent on the *p.d.* across it; it is a *non-linear* resistor

voltmeter   An instrument for the measurement of *voltage*

watt   The unit of electrical *power*, symbol W

wattmeter   An instrument for measuring electrical *power*

weber   The unit of *magnetic flux*, symbol Wb

Wheatstone bridge   A circuit for the measurement of *resistance*

zener diode   A *diode* with a well-defined reverse breakdown voltage, which can be operated in the reverse breakdown mode

PRINCIPLES OF ELECTRICITY

1.1 ATOMIC STRUCTURE

About 100 basic substances or **chemical elements** are known to man, each element consisting of a number of smaller parts known as **atoms**. Each atom comprises several much smaller particles, the principle ones being **electrons, protons** and **neutrons**.

The difference between the smaller particles lies not only in their difference in mass (a proton is 1840 times more 'massive' than an electron), but also in the **electrical charge** associated with them. For example, a proton has a *positive electrical charge* whilst an electron has a *negative electrical charge*; the charge on the proton is equal to but of opposite polarity to that on the electron. The electrical charge on either the proton or the electron is very small, in fact it is so small that one ampere of current is associated with the movement of over *six million billion electrons per second*.

The mass of the neutron is equal to that of the proton, but it has no electrical charge. In the latter respect it has little use in electrical circuits.

The nature of matter ensures that each atom is electrically balanced, that **it has as many electrons as it has protons**. Under certain circumstances an atom, or a molecule, or a group of atoms can acquire an electrical charge; the atom or group of atoms is then known as an **ion**. A **negative ion** (an **anion**) contains more electrons than are necessary for electrical neutrality; a **positive ion** (a **cation**) contains fewer electrons than necessary for neutrality.

The protons and neutrons are concentrated in the centre or **nucleus** of the atom as shown in Figure 1.1. The electrons *orbit around the nucleus* in what are known as **layers**, or **energy bands** or **shells**. A simple analogy of an atom is that of a multi-storey car park. The ground level, or 'zero energy' level can be regarded as the nucleus of the atom, whilst the higher found. The ground level is filled with 'car parking for staff cars' which we

fig 1.1 *electrons in orbit around a nucleus*

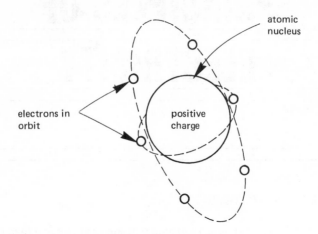

will regard as protons and neutrons. As other people come along to park their cars, they must do so in the 'higher energy' levels. So it is with atoms – the lower shells are filled with electrons before the higher shells.

The electrons which take part in the process of electrical conduction are in the outermost shell or highest energy level shell of the electron; this is known as the **valence shell** or **valence energy band**. For an electron to take part in electrical conduction, it must be free to 'move' within its energy band. In the multi-storey car park analogy this is equivalent to the cars on the uppermost floor having the most room to move about.

1.2 ELECTRONIC 'HOLES'

The application of an electrical voltage to a conductor results in electrons in the outermost shell (the valence shell) being subjected to an electrical force. This force tries to propel the electrons towards the positive pole of the supply; if the force is sufficiently great, some electrons escape from the forces which bind them to the atom. The electrons which arrive at the positive pole of the battery constitute flow of **electrical current**.

However, when an atom loses an electron its electrical neutrality is lost, and the remainder of the atom takes on a net positive charge. This positive charge will attract any mobile electron in its vicinity; in this way, when an electron moves from one part of the conductor to another, it leaves behind it a resulting positive charge (arising from the loss of the electron at that point). Thus, as the electron 'moves' in one direction, a positive charge 'moves' in the opposite direction. On this basis, it is possible to describe the *mobile positive charge* as a **hole** into which any electron can fall (a 'hole' may be thought of as the absence of an electron where one would normally be found).

Electrical engineers therefore think of a *mobile negative charge carrier* as an *electron*, and a *mobile positive charge carrier* as a *hole*.

1.3 CONDUCTORS, SEMICONDUCTORS AND INSULATORS

A **conductor** is an electrical material (usually a metal) which offers very little resistance to electrical current. The reason that certain materials are good conductors is that the outer orbits (the valence shells) in adjacent atoms overlap one another, allowing electrons to move freely between the atoms.

An **insulator** (such as glass or plastic) offers a very high resistance to current flow. The reason that some materials are good insulators is that the outer orbits of the atoms do not overlap one another, making it very difficult for electrons to move through the material.

A **semiconductor** is a material whose resistance is midway between that of a good conductor and that of a good insulator. Other properties are involved in the selection of a semiconductor material for electrical and electronic purposes; these properties are discussed later in the book. Commonly used semiconductor materials include silicon and germanium (in diodes, transistors and integrated circuits), cadmium sulphide (in photoconductive cells), gallium arsenide (in lasers, and light-emitting diodes), etc. Silicon is the most widely used material, and is found in many rocks and stones (sand is silicon dioxide).

1.4 VOLTAGE AND CURRENT

Voltage is the electrical equivalent of mechanical potential. If a person drops a rock from the first storey of a building, the velocity that the rock attains on reaching the ground is fairly small. However, if the rock is taken to the twentieth floor of the building, it has a much greater potential energy and, when it is dropped it reaches a much higher velocity on reaching the ground. The *potential energy* of an electrical supply is given by its **voltage** and the greater the voltage of the supply source, the greater its potential to produce **electrical current** in any given circuit connected to its terminals (this is analogous to the velocity of the rock in the mechanical case). Thus the potential of a 240-volt supply to produce current is twenty times that of a 12-volt supply.

The electrical potential between two points in a circuit is known as the **potential difference** or **p.d.** between the points. A battery or electrical generator has the ability to produce current flow in a circuit, the voltage which produces the current being known as the **electromotive force**

(e.m.f.). The term electromotive force strictly applies to the source of electrical energy, but is sometimes (incorrectly) confused with potential difference. Potential difference and e.m.f. are both measured in **volts**, symbol V.

The **current** in a circuit is due to the movement of charge carriers through the circuit. The charge carriers may be either electrons (negative charge carriers) or holes (positive charge carriers), or both. Unless stated to the contrary, we will assume *conventional current flow* in electrical circuit, that is we assume that **current is due to the movement of positive charge carriers (holes) which leave the positive terminal of the supply source and return to the negative terminal.** The current in an electrical circuit is measured in **amperes**, symbol A, and is sometimes (incorrectly) referred to as 'amps'.

A simple electrical circuit comprising a battery of e.m.f. 10 V which is connected to a heater of fixed resistance is shown in Figure 1.2(a); let us suppose that the current drawn by the heater is 1 A. If two 10-V batteries are connected in series with one another, as shown in Figure 1.2(b), the

fig 1.2 *relationship between the voltage and current in a circuit of constant resistance*

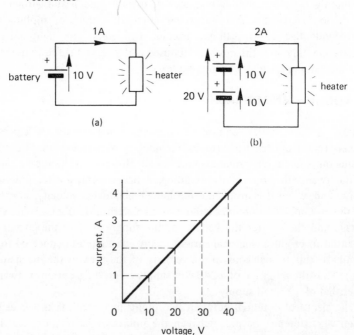

e.m.f. in the circuit is doubled at 20 V; the net result is that the current in the circuit is also doubled. If the e.m.f. is increased to 30 V, the current is increased to 3 A, and so on.

A graph showing the relationship between the e.m.f. in the circuit and the current is shown in Figure 1.2(c), and is seen to be a straight line passing through the origin; that is, the current is zero when the supply voltage is zero. This relationship is summed up by Ohm's law in section 1.5.

1.5 OHM'S LAW

The graph in Figure 1.2(c) giving the relationship between the voltage across an element of fixed resistance and the current through it shows that the **current is proportional to the applied voltage**. This relationship (attributed to the German teacher G. S. Ohm) is given the name **Ohm's law**, and is stated below

$$\text{Current, } I = \frac{\text{e.m.f. } (E) \text{ or voltage } (V)}{\text{resistance, } R \text{ (ohms)}} \text{ amperes, A}$$

where I is the current in amperes flowing in the circuit of *resistance R ohms* (given the Greek symbol Ω [omega]). A ghoulish way of remembering Ohm's law is:

Interment (I) = Earth (E) over Remains (R)

Alternatively, Ohm's law may be stated in one of the following ways:

E or $V = IR$ (volts) (V)

$$R = \frac{E \text{ or } V}{I} \text{ (ohms) } (\Omega)$$

The Greek symbol Ω (omega) was first used by W. H. Preece to represent the units of resistance in a lecture to Indian Telegraph Service cadets at the Hartley College (now the University, Southampton) in 1867.

The resistance of a conductor (or of an insulator) indicates the ability of the circuit to resist the flow of electricity; a low resistance, for example, $0.001 \ \Omega$, implies that the element is a poor resistor of electricity (or a good conductor!), whilst a high value, for example, $100\,000\,000 \ \Omega$ implies that the element is a very good resistor (or a poor conductor).

Ohm's law is illustrated in the following simple examples. If a current of 0.5 A flows in a circuit of resistance 15 Ω, the p.d. across the circuit is

voltage = IR = 0.5 × 15 = 7.5 V

Also, if an e.m.f. of 10 V is applied to a circuit of resistance 10 000 Ω, the current in the circuit is

$$I = \frac{E}{R} = \frac{10}{10\,000} = 0.0001 \text{ A}$$

You should note that the resistance of a conductor may vary with temperature, and the effect of temperature on resistance is studied in detail in Chapter 3.

1.6 CONDUCTANCE

In some cases it is convenient to use the reciprocal of resistance, that is $\frac{1}{R}$ rather than resistance itself. This reciprocal is known as the **conductance**, symbol G, of the conductor; this value indicates the ability of the circuit to *conduct* electricity. The unit of conductance is the **siemens** (symbol S), and

$$\text{conductance, } G = \frac{1}{R} \text{ S}$$

A very low value of conductance, e.g., 0.000 000 0001 S implies that the circuit is a poor conductor of electricity (and is a good insulator), whilst a high value of conductance, e.g., 1000 S implies that it is a good conductor (or poor resistor). For example, a conductor of resistance 0.001 Ω has a conductance of

$$G = \frac{1}{R} = 1/0.001 = 1000 \text{ S}$$

whilst an insulator of resistance 10 000 000 Ω has a conductance of

$$G = \frac{1}{R} = \frac{1}{10\,000\,000} = 0.000\,000\,1 \text{ S}$$

1.7 LINEAR AND NON-LINEAR RESISTORS

The majority of metals are good conductos and the current–voltage relationship for these materials obeys Ohm's law; that is, the I–V graph passes through the origin (see Figure 1.3(a)), and an increase in voltage (either positive or negative) produces a proportional change in the current. That is to say

$$I \, \alpha \, V$$

where the Greek symbol α (alpha) means 'is proportional to'. In such a circuit element, doubling the voltage across the element has the effect of doubling the current through it.

fig. 1.3 *the linear resistor (or conductor) in (a) obeys Ohm's law; the circuit elements having the characteristics shown in (b), (c) and (d) do not obey Ohm's law*

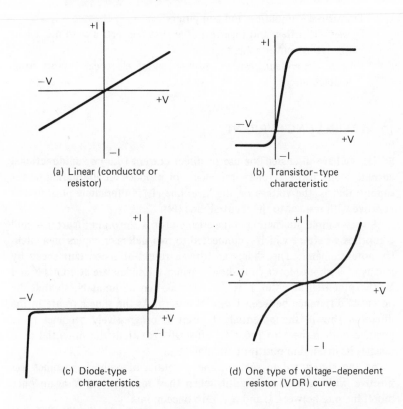

(a) Linear (conductor or resistor)

(b) Transistor-type characteristic

(c) Diode-type characteristics

(d) One type of voltage-dependent resistor (VDR) curve

However, there are other circuit elements which do not obey Ohm's law. These are known as **non-linear resistors** or **non-linear conductors**. The characteristics in diagrams (b) to (d) correspond to practical non-linear devices.

In the case of a *transistor* (Figure 1.3(b), the current increases rapidly from the origin of the graph but, when the p.d. across it is about 0.6 to 1.0 V, the current becomes more-or-less constant; the value of the current depends whether the voltage is positive or negative. In the case of a *semiconductor diode* (Figure 1.3(c)), the current rises very rapidly for positive voltages; for negative voltages the current is practically zero up to a fairly high voltage, at which point the current increases rapidly (this is known as reverse breakdown). In the case of a *voltage dependent resistor*, VDR (Figure 1.3(d)), the current through it increases at a progressively rapid rate as the voltage across it increases.

Although the *I-V* characteristic of the devices in Figure 1.3 may not obey Ohm's law, each finds its own special applications in electrical engineering. The following applications are typical:

Transistors – amplifiers and computers

Diodes – rectifiers and invertors (the first converts a.c. to d.c. while the latter converts d.c. to a.c.)

Voltage dependent resistor – protection of electrical circuits from voltage surges.

1.8 ALTERNATING CURRENT

So far we have discussed the use of **direct current** (d.c.) or **unidirectional current**. Many forms of electrical generator produce an **alternating** power supply; that is, the voltage on the 'live' line (L) is alternately positive and negative with respect to the 'neutral' line (N).

A very simple alternating voltage generator is shown in Figure 1.4, and comprises a battery which is connected to two **sliders** or **wipers** on a circular potentiometer. The sliders are driven round at a constant speed by means of a *prime mover* (not shown. When the sliders are at points X and Y, the potential at point L is the same as that at point N, so that the potential difference between L and N is zero. As the sliders rotate in the direction shown, the potential of point L progressively becomes more positive with respect to point N; after $90°$ rotation, the potential at L reaches its maximum positive potential $+E_m$.

As the sliders rotate further, the potential of point L becomes less positive with respect to N until, when they reach X and Y again once more, the p.d. between L and N is zero once more.

Further rotation causes point L to become negative with respect to point N. When the slider connected to the positive pole of the battery reaches N, point L is at its maximum negative potential, $-E_m$. Further rotation causes point L to become less negative with respect to point N until, after $360°$ rotation from the start of the waveform, the sliders reach their original position. Consequently, the potential of point L with respect to point N *alternates* about zero, as shown in the waveform in Figure 1.4.

For safety reasons, many electrical circuits have one point of the electrical circuit connected to earth potential. In the case of Figure 1.4 this is the *N*-line or **neutral line**; the other line (line *L*) is known as the **live line**.

One cycle of an alternating electrical supply is shown in the waveform in Figure 1.4, and commences at point O and finishes at the start of the following cycle. This corresponds to a rotation of $360°$ or 2π radians.

The time taken to complete one cycle of the supply frequency is

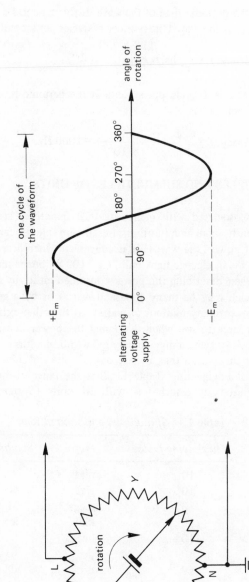

fig 1.4 a simple alternating current generator

known as the **periodic time** of the wave. If this time is T seconds, then the **frequency**, f, of the wave (the number of cycles per second) is given by

$$\text{frequency}, f = \frac{1}{T} \text{ hertz (Hz)}$$

where 1 hertz is 1 cycle per second. If the periodic time of the wave is 1 ms, the frequency is

$$\text{frequency}, f = \frac{1}{1 \text{ ms}} = \frac{1}{1 \times 10^{-3}} = 1000 \text{ Hz}.$$

1.9 MULTIPLES AND SUBMULTIPLES OF UNITS

The units associated with many practical quantities are in some cases inconveniently small and, in other cases, inconveniently large. For example the unit of power, the watt (W), is convenient when describing the power consumed by an electric light bulb, e.g., 100 W, but is much too small to be used when describing the power consumed either by a large electrical heater (which may be many thousand watts) or by a generating station (which may be many millions of watts). At the other extreme, the watt is much too large to use when discussing the power consumed by a low-power transistor (which may be only a few thousandths of a watt).

It is for this reason that a range of multiples and submultiples are used in Electrical Engineering. Table 1.1 lists the range of multiples and sub-multiples used in association with SI units (*Système International*

Table 1.1 *SI multiples and submultiples*

Multiplying factor	Prefix	Symbol
10^{12}	tera	T
10^9	giga	G
10^6	mega	M
10^3	kilo	k
10^2	hecto	h
10	deca	da
10^{-1}	deci	d
10^{-2}	centi	c
10^{-3}	milli	m
10^{-6}	micro	μ
10^{-9}	nano	n
10^{-12}	pico	p
10^{-15}	femto	f
10^{-18}	atto	a

d'Unités), which is the system of units adopted by Electrical and Electronic Engineers.

For example, 11 000 000 watts could be written either as 11×10^6 W or as 11 MW; 0.0001 W could be written either as 10^{-4} W or as 0.1 mW or as 100 μW. Strictly speaking, any of the prefixes in Table 1.1 can be used but, in practice, deca- and hecto- are practically never used; femto- and atto- refer to very small values indeed, and have only a limited use.

You should also note that compound prefixes are avoided wherever possible; for example, the multiple 10^{-9} is described as nano- rather than milli-micro. That is, 10^{-9} metre (m) is described as 1 nm and not as 1 m μm.

1.10 SOME BASIC ELECTRICAL QUANTITIES

Some of the more commonly used units in electrical and electronic engineering are briefly described below.

Electrical quantity (symbol Q)
The quantity of electricity passing a point in an electrical circuit is

Quantity, $Q = I$ (amperes) $\times t$ (seconds) coulombs (C)

For example, if a circuit carries a current of 5 amperes for 15 seconds, the quantity of electricity is

Quantity, $Q = It = 5 \times 15 = 75$ C

Electrical power (symbol P)
Power is the rate of dissipation of energy and is given by

Power, $P = E$ (volts) $\times I$ (amperes)   watts (W)

If the electrical voltage across a resistive circuit element is 240 V, and if the current passing through it is 13 A, the power consumed is

Power, $P = EI = 240 \times 13 = 3120$ W

Also from Ohm's law, $E = IR$, hence

Power, $P = EI = (IR) \times I = I^2R$ W

Since $I = \frac{E}{R}$, then

Power, $P = EI = E \times \dfrac{E}{R} = \dfrac{E^2}{R}$ W

Electrical energy (symbol W)

The energy dissipated in an electrical circuit is given by

Energy, $W = E$ (volts) $\times I$ (amperes) $\times t$ (seconds) joules (J)

If the p.d. across a circuit element is 100 V and it carries a current of 10 A for 15 s, the energy consumed is

Energy, $W = EIt = 100 \times 10 \times 15 = 15\,000$ J or watt-seconds.

The **commercial unit of electrical energy** is the **kilowatt-hour** or kWh, which corresponds to 1 kilowatt of power being consumed continuously for a period of one hour.

SELF-TEST QUESTIONS

1. What is the difference btween an atom and a chemical element? How do electrons, protons and neutrons differ?
2. Copper is a conductor and glass is an insulator. Why do these materials have different electrical resistance? Why is a 'semiconductor' so named?
3. Explain the terms 'voltage' and 'potential drop'.
4. Using Ohm's law, calculate the current flowing in a resistance of 100 Ω which has a p.d. of 20 V across it.
5. Calculate the conductance in siemens of a circuit which carries a current of 100 A, the p.d. across the circuit being 10 V.
6. The following values of current were measured in a circuit when the voltage across the circuit was (i) 10 V, (ii) 20 V, (iii) 30 V: 10 A, 40 A, 150 A. Plot the $I-V$ characteristic of the circuit and state if it is linear or non-linear.
7. An alternating current waveform has a periodic time of 1 ms, calculate the frequency of the current. What is the periodic time of an alternating voltage of frequency 100 kHz?
8. A current of 20 A flows in a circuit for 20 s, the voltage applied to the circuit being 20 V. Calculate the electrical quantity, the power and the energy consumed.

SUMMARY OF IMPORTANT FACTS

A chemical **element** is built up from **atoms**. Each atom is made up of **electrons, protons** and **neutrons**. The electron has a **negative charge** and the proton a **positive charge**; the two electrical charges are **equal and opposite**, but the mass of the proton is **1840 times greater** than that of the electron.

Current flow is a combination of **electron flow** (negative charge carriers) and **hole flow** (positive charge carriers).

A **conductor** offers low resistance to current flow, an **insulator** offers high resistance. The resistance of a **semiconductor** is mid-way between the two extremes.

Voltage or **e.m.f.** is a measure of the potential of an electric circuit to produce current flow.

Ohm's law states that

$$E = IR$$

$$I = \frac{E}{R}$$

$$R = \frac{E}{I}$$

where E is in volts, I in amperes and R in ohms.

The **conductance** (G) of an electrical circuit is measured in siemens (S) and $G = \frac{1}{R}$ (R in ohms).

A **linear circuit element** is one whose resistance is constant despite fluctuations of voltage and current. The resistance of a **non-linear circuit element** varies with fluctuations of voltage and current.

A **direct current** or **unidirectional current** always flows in the same direction around a circuit. An **alternating current** periodically reverses its direction of flow. If the **periodic time** of an alternating wave is T seconds, the **frequency** of the alternating wave is $f = \frac{1}{T}$ Hz.

CHAPTER 2

ELECTROCHEMISTRY, BATTERIES AND OTHER SOURCES OF e.m.f.

2.1 ELECTROCHEMICAL EFFECT

The chemical effect of an electric current is the basis of the *electroplating industry*; the flow of electric current between two **electrodes** (one being known as the **anode** and the other as the **cathode**) in a liquid (the **electrolyte**) causes material to be lost from one of the electrodes and deposited on the other.

The converse is true, that is, chemical action can produce an e.m.f. (for example, in an electric battery).

All these electrochemical effects depend on the electrolyte. The majority of **pure liquids** are good insulators (for example, *pure water* is a good insulator), but **liquids containing salts** will conduct electricity. You should also note that some liquids such as mercury (which is a liquid metal) are good conductors.

2.2 IONS

An atom has a nucleus (positively charged) surrounded by electrons (negatively charged) which orbit around the nucleus in shells or layers. In an electrically neutral atom, the positive and negative charges are equal to one another and cancel out. However, the electrons in the outermost orbit (the valence electrons) are fairly loosely attracted to the parent atom and can easily be detached. In fact, it is possible for either a chemical reaction or an electric field to cause an atom either to lose an electron or to gain one.

When an atom loses an electron, its charge balance is upset and the parent atom (which has lost the negatively charged electron) is left with a charge of +1 unit of electricity (equivalent to the charge on a proton). In this case the parent atom is described as a **positive ion** or a **cation**; *the ion retains the characteristics of the original element because the nucleus* *remains intact.* When an atom gains an electron, it has a net charge of -1 unit of electricity (equivalent to the charge on one electron) and is described as a **negative ion** or **anion**.

Since like charges attract and unlike charges repel, a *negative ion is attracted to a positively charged electrode* and is *repelled by a negatively charged electrode* (and vice versa for a positive ion). For this reason, current flow can take place in ionised material as follows: in liquids such as electrolytes and in gases (for instance, in fluorescent tubes) current flow is due to the movement of ions between two oppositely charged electrodes, the current flow being known as **ionisation current**.

Ions are formed in a liquid either when a salt or an acid is dissolved in it; additionally, when a metal is immersed either in an alkaline or an acid solution, ionisation occurs. The resulting liquid is a fairly good conductor of electricity and is known as an **electrolyte**.

2.3 ELECTROLYSIS

Electrolysis is the process of decomposing an electrolyte by passing an electric current through it. The chemical action is seen at the **electrodes** (where the current enters or leaves the electrolyte in which the electrodes are immersed – see Figure 2.1). The electrode by which the current enters the electrolyte is known as the **anode**; the current leaves via the **cathode** electrode.

Electrolysis is the basis of the **electroplating industry** in which one metal is plated with another metal. It is also used in the **extraction** and **refining** industry; copper, zinc and aluminium being examples of metals extracted by this means. It is also the basis of other **electrolytic processes** involved in the production of certain types of gas; for example, hydrogen

fig 2.1 *electrolysis*

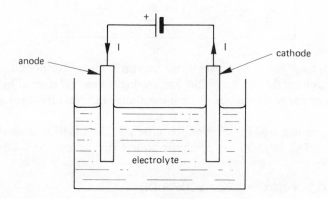

and oxygen can be produced by breaking down water into its basic chemical constituents.

2.4 AN EXAMPLE OF ELECTROLYSIS

To simplify the representation of ions in chemical processes, a form of shorthand is used to denote different types of ion. Positive ions have a positive sign associated with them in the form of a superscript, for example, a hydrogen ion is shown as H^+. Each 'unit' positive charge associated with the ion is shown as a 'plus'; thus Cu^{++} means that the copper ion is **doubly charged**. Negative ions have a negative subscript, for example, the sulphate ion SO_4^{--} is a doubly-charged anion.

Consider the example in Figure 2.2 in which sulphuric acid is added to water to form dilute sulphuric acid. The sulphuric acid molecule, H_2SO_4, splits in the water into two hydrogen cations ($2H^+$) and a sulphate anion (SO_4^{--}). When the two electrodes in the solution are connected to the battery, the sulphate ions (negative ions) are attracted to the anode and the hydrogen ions (positive ions) to the cathode.

fig 2.2 *production of hydrogen and oxygen*

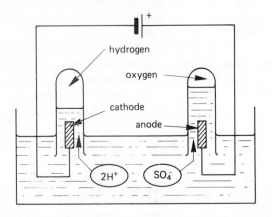

When a hydrogen ion arrives at the negatively charged cathode it picks up an electron; this discharges the ion, leaving the neutral atom of hydrogen gas to rise to the surface where it is collected in a test tube (see Figure 2.2).

The reaction when a sulphate ion arrives at the anode is rather more complex. The sulphate ion combines with water to give the following reaction:

$$SO_4^{--} + 2H_2O = SO_4^{--} + 2(H^+ + OH^-)$$

$$= (2H^+ + SO_4^{--}) + 2OH^-$$
$$= H_2SO_4 + 2OH^-$$

The first line shows that the water dissociates into H^+ and OH^- ions, which further combine to produce sulphuric acid (H_2SO_4) and OH^- ions. The OH^- ions rearrange themselves into water and oxygen as follows:

$$2OH^- = H_2O + O^{--}$$

The negative charge associated with the O^{--} is discharged at the anode, the resulting neutral atoms of oxygen rise to the surface and are collected in the test tube.

From this, you will see that the sulphuric acid is constantly being replaced by the process and is not diminished, but the water is converted into its basic elements, namely, hydrogen and oxygen gas. Clearly, unless the water is replenished, the concentration of the acid increases slowly with time. Since the amount of acid is constant, it can be regarded as a **catalyst**.

2.5 ELECTROPLATING

When two different metals are immersed in a solution which is a salt *of the same material as the anode*, and a current is passed between the two metals, metal ions from the anode dissolve into the solution and, at the same time, ions of the anode material are deposited on the cathode.

Copper plating
If the anode is of copper and the electrolyte is copper sulphate, the cathode (of, say, iron) is plated with copper.

Silver plating
A silver anode is used and the electrolyte is cyanide of silver; the cathode must be a metal which does not chemically react with the electrolyte.

2.6 FARADAY'S LAWS OF ELECTROLYSIS

In the process of electrolysis, ions migrate towards the electrodes and, on arrival at the electrodes, give up some electrical charge. The charge given up *per ion* is

charge per ion = ze coulombs

where z is the number of valence electrons and e is the charge per electron. Hence the total charge Q given up by N ions is

$$Q = Nze \text{ coulombs} \tag{2.1}$$

or

$$\text{number of ions, } N = \frac{Q}{ze}$$

Michael Faraday discovered this relationship in 1833, and stated two laws which govern electrolysis as follows:

First law:

> *The mass of material deposited (or gas released) is proportional to the quantity of electricity (current × time) which passes through the electrolyte.*

(This is suggested by Eqn (2.1).)

Second law:

> *The mass of material in grammes deposited by the passage of 26.8 ampere hours of electricity is equal to the chemical equivalent of that material.*

Example

When an electrical current passes through a solution of sodium chloride in water, sodium hydroxide and hydrochloric acid are produced. If the chemical equivalent of sodium hydroxide is 40 and that of hydrochloric acid is 36.5, determine how much hydrochloric acid and sodium hydroxide are produced when 100 A flows for 10 mins.

Solution

$$I = 100 \text{ A}, t = 10 \text{ mins} = \frac{1}{6} \text{ hour}$$

$$Q = It = 100 \times \frac{1}{6} = 16.67 \text{ Ah}$$

From Faraday's second law of electrolysis:

mass of material deposited

$$= \text{chemical equivalent} \times \frac{\text{quantity of electricity in Ah}}{26.8}$$

hence

$$\text{mass of sodium hydroxide} = 40 \times \frac{16.67}{26.8} = 24.88 \text{ g (Ans.)}$$

and

$$\text{mass of hydrochloric acid} = 36.5 \times \frac{16.67}{26.8} = 22.7 \text{ g (Ans.)}$$

2.7 CELLS AND BATTERIES

A **cell** contains two **plates** immersed in an **electrolyte**, the resulting chemical action in the cell producing an e.m.f. between the plates. Cells can be grouped into two categories. A **primary cell** cannot be recharged and, after the cell is 'spent' it must be discarded (this is because the chemical action inside the cell cannot be 'reversed'). A **secondary cell** or **storage cell** can be recharged because the chemical action inside it is reversed when a 'charging' current is passed through it. Table 2.1 lists the e.m.f. associated with a number of popular cells.

Cells are also subdivided into 'dry' cells and 'wet' cells. A **dry cell** is one which has a *moist electrolyte*, allowing it to be used in any physical position (an electric torch cell is an example). A **wet cell** is one which has a *liquid electrolyte* which will spill if the cell is turned upside down (a cell in a conventional lead-acid auto battery is an example). There is, of course, a range of sealed rechargeable cells which are capable of being discharged or charged in any position; the electrolyte in these cells cannot be replaced.

Table 2.1 *Cell voltages*

Type of cell	Category	e.m.f.
Carbon zinc	primary	1.5
mercury oxide	primary	1.35
silver oxide	primary	1.5
zinc chloride	primary	1.5
manganese dioxide	primary or secondary	1.5
lead acid	secondary	2.1
nickel cadmium	secondary	1.35
nickel iron	secondary	1.25
silver cadmium	secondary	1.1
silver zinc	secondary	1.5

A **battery** is an interconnected group of cells (usually connected in series) to provide either a higher voltage and/or a higher current than can be obtained from one cell.

2.8 A SIMPLE VOLTAIC CELL

A simple **voltaic cell** is shown in Figure 2.3, the e.m.f. produced by the cell being given by the algebraic difference between the 'contact potential' e.m.fs of the two electrodes; the contact potential of commonly used materials are listed in Table 2.2.

In the cell in Figure 2.3, zinc reacts with the acid in the electrolyte, the electrons released in the process giving the zinc electrode a negative potential.

The copper electrode does, to some extent, react with the acid but to a far lesser degree than does the zinc. The sulphuric acid in the electrolyte dissociates into its negative sulphate ions, SO_4^{--}, and its positive ions, $2H^+$. The negative ions give up their charge to the zinc electrode and the positive ions move to the copper electrode where they give up their charge, causing the copper plate to be positive. Once the hydrogen ions give up their charge, they can be seen as hydrogen bubbles on the surface of the copper electrode; they can then float to the surface to be absorbed into the atmosphere.

fig 2.3 *a simple voltaic cell*

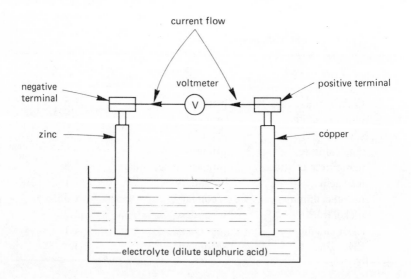

The e.m.f. produced by the cell can be deduced from the electro-chemical series in Table 2.2. The e.m.f. of the copper-zinc cell is *the difference in potential between the elements*. In this case it is

$$0.34 - (-0.76) = 1.1 \text{ V}.$$

Table 2.2 *Electrode potentials*

Element	Potential (V)
magnesium	-2.40
aluminium	-1.70
zinc	-0.76
cadmium	-0.40
nickel	-0.23
lead	-0.13
hydrogen	0.00
copper	0.34
mercury	0.80
gold	1.50

2.9 INTERNAL RESISTANCE OF A CELL

A cell produces a theoretically 'ideal' e.m.f. E (see Figure 2.4) which can be predicted from the table of electrochemical elements. If you measure the p.d. between the terminals of the cell **when the load R is disconnected** (that is when the load current is zero) you are measuring the e.m.f., E; this voltage is known as the **no-load terminal voltage of the cell**.

When the load resistance, R, is connected, a current flows through the cell; that is, it flows not only through the electrolyte but also through the contact between the electrolyte and the electrodes. Since, at normal temperature, no material is resistanceless, the cell has an **internal resistance**, r (see Figure 2.4), which is in series with the e.m.f., E.

The potential drop in the internal resistance r (known as the *internal voltage drop*) of the cell causes the **terminal voltage**, V_T, under 'loaded' conditions to be less than the no-load terminal voltage, E. The equation for the terminal voltage is:

Terminal voltage = e.m.f. − internal voltage drop

or

$$V_T = E - Ir$$

fig 2.4 *equivalent electrical circuit of a cell*

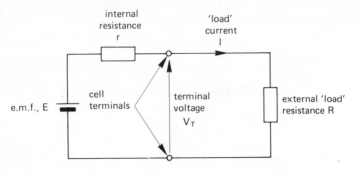

Example

Calculate (a) the current flowing from and (b) the terminal voltage of a cell of e.m.f. 1.5 V and internal resistance 2 Ω when the load resistance is (i) 100 Ω, (ii) 10 Ω, (iii) 2 Ω, (iv) zero (that is, when the terminals of the cell are shorted together).

Solution

Since $E = 1.5$ V, $r = 2$ Ω, the current, I, drawn by the load resistor R (see Figure 2.4) is given by

$$I = \frac{1.5}{(2 + R)} \text{ A}$$

The terminal voltage, V_T, is calculated from the expression

$$V_T = IR \text{ V}$$

An example calculation is given below for $R = 100$ Ω.

(a) $I = \dfrac{1.5}{(2 + R)} = \dfrac{1.5}{(2 + 100)} = 0.0147 \text{ A (Ans.)}$

and

(b) $V_T = IR = 0.0147 \times 100 = 1.47 \text{ V (Ans.)}$

The above calculation shows that the internal resistance has caused an 'internal' voltage drop of $1.5 - 1.47 = 0.03$ V. The results for the remainder of the calculation are listed below, and it would be an interesting exercise for the reader to verify them.

R (ohm)	100	10	2	0
I (amperes)	0.0147	0.125	0.375	0.75
V_T (volts)	1.47	1.25	0.75	0

In the case where the cell terminals are short-circuited ($R = 0$), the terminal voltage is zero because the load resistance is zero. What, you may ask, has happened to the e.m.f.? The explanation is that the p.d. across the internal resistance r is

$$Ir = I \times r = 0.75 \times 2 = 1.5 \text{ V}$$

In this case, the whole of the e.m.f. is 'dropped' across the internal resistance of the cell. That is:

p.d. across load $= E - $ 'internal voltage drop'

$$= 1.5 - 1.5 = 0 \text{ V}$$

leaving zero volts between the terminals of the cell. **You are advised against the practice of short-circuiting the terminals of a cell because of the risk of the cell exploding as a result of the intense chemical activity in the cell.**

2.10 LIMITATIONS OF SIMPLE CELLS

In practice the terminal voltage of a cell is not only less than the e.m.f. of the cell but, with use, it falls to an even lower value. The principal reason for the latter is **polarisation** of one of the electrodes, and is due to hydrogen gas bubbles collecting on its surface.

Consider the copper-zinc cell in Figure 2.3. In the process of producing a voltage, hydrogen gas bubbles form on the surface of the copper electrode. If all the surface of the copper was covered with hydrogen bubbles, the cell would become a hydrogen-zinc cell having a theoretical e.m.f. of

$$0.00 - (-0.76) = 0.76 \text{ V}$$

(see also Table 2.2). Moreover, since hydrogen is a poor conductor of electricity, the internal resistance of the cell would be very high. This has a further effect of reducing the terminal voltage when current is drawn from it.

The effects of polarisation are overcome by adding a chemical to the electrode; the function of the chemical is to combine with the hydrogen to form water. Such a chemical is described as a **depolarising agent** or **depolariser**. In some cases, powdered carbon is added to the depolariser to improve its conductivity and therefore reduce the internal resistance of the cell.

Another problem with cells using a zinc electrode is known as **local action**. Local action is due to the difficulty of obtaining zinc which is free from impurities such as iron or lead. If the impurity is on the surface of the zinc, it form a 'local' cell which, because of the intimate contact between the impurity and the zinc, is short-circuited. Because of the short-circuiting action, each local cell is almost instantaneously discharged, the energy appearing as heat in the cell; the action also causes some of the zinc to be consumed in the process. A remedy is to use an **amalgamated zinc electrode**, which is zinc which has been treated with mercury; the impurities normally in the zinc are not soluble in the mercury and so do not present a problem.

2.11 THE 'DRY' CELL

Early 'dry' cells were versions of the Leclanché 'wet' cell, in which the positive electrode was a rod of carbon which passed into a porous pot packed with a mixture of powdered carbon and manganese dioxide, the latter being a depolarising agent.

The powdered carbon inside the porous pot effectively increased the volume of the positive electrode; the function of the porous pot was simply to keep the mixture in close contact with the carbon rod. The porous pot stood in a glass vessel containing the zinc (negative) electrode together with the electrolyte of dilute ammonium chloride. This type of cell had an almost indefinite life, the maintenance consisting of the following:

    1. the zinc rod needed to be replaced from time-to-time;
    2. some ammonium chloride needed to be added occasionally;
    3. the electrolyte needed 'topping up' with tap water from time-to-time.

A simplified cross-section of a modern dry cell is shown in Figure 2.5.

fig 2.5 *the construction of a 'dry' cell*

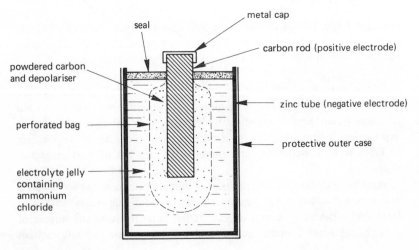

The positive electrode (the **anode**) is a carbon rod (as it was in the Leclanché cell), which is in intimate contact with a mixture of powdered carbon and an depolariser which are contained in a perforated bag. The perforations in the bag allow the electrolyte to come into contact with the powdered carbon. The electrolyte is in the form of a jelly, so that the cell can be used in any physical position. The negative electrode (the **cathode**) is a zinc tube which provides a rigid case for the cell.

Different types of dry cell are manufactured for a range of applications. For example, some cells are specifically designed for use with calculators, some for use with clocks, and others for use with torches, etc. Two of the features allowing a specific performance to be achieved are the number and size of the holes in the perforated bag which holds the depolariser; these affect the rate of discharge of current from the cell.

2.12 OTHER TYPES OF PRIMARY CELL

The zinc chloride cell
This is generally similar to the dry cell described in section 2.11, but the electrolyte contains zinc chloride.

The mercury cell
The anode in this type of cell is made of zinc, the cathode being a mercury compound, and the electrolyte is sodium or potassium hydroxide. These cells are manufactured in cylindrical and 'button' shapes, and they maintain a fairly constant voltage over their working life. The terminal voltage is typically 1.35–1.4 V per cell.

The silver oxide cell
The anode in this cell is made of zinc, the cathode is silver oxide, and the electrolyte is potassium or sodium hydroxide. The button-shape cells give a terminal voltage of about 1.5 V and are widely used in many popular applications including electronic watches, hearing aids and cameras.

The lithium cell
This type of cell can provide about ten times more energy than an equivalent size of carbon–zinc cell but, since lithium is a chemically active element, there may be a risk of explosion without warning. Depending on the electrolyte, the e.m.f. is in the range 2.9 to 3.7 V per cell.

2.13 STORAGE BATTERIES

Rechargable cells are often connected in series to form a **storage battery**, a *car battery* being an example; a storage battery is frequently called an

accumulator. The cells of the battery have a reversible chemical action and, when current is passed through them in the 'reverse' direction (when compared with the discharging state), the original material of the electrodes is re-formed. This allows the battery to be repeatedly discharged and charged. The chemical action is rather complex, and one of the most important of them – the lead-acid battery – is described here.

The lead–acid accumulator

A simplified cross-section through a lead-acid battery is shown in Figure 2.6. The accumulator has as many positive plates as it has negative plates, the **separators** between the plates being porous insulators. The accumulator is usually houses in a glass or plastic vessel containing the electrolyte which is dilute sulphuric acid.

Different types of chemical activity occur at each of the two plates both during the electrical charging process and during its discharge. However, the process can be summarised and simplified by the following equation:

condition when charged				condition when discharged	
positive plate	negative plate	electrolyte		both plates	electrolyte
PbO_2 +	Pb +	$2H_2SO_4$	$\rightleftharpoons$	$2PbSO_4$ +	$2H_2O$

Both plates are a lead alloy grid, the holes in the grid being packed with a chemical substance. In the case of the **positive plate** of a charged accumulator (see the above equation), the grids contain lead peroxide (PbO_2) and,

fig 2.6 *cross-section through a lead-acid battery*

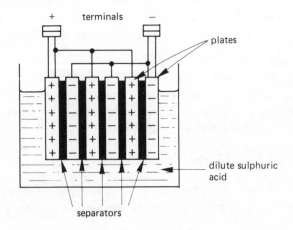

in the case of the **negative plate**, it is spongy lead (Pb). During the charging process of the battery, the **specific gravity**, that is, the strength of the electrolyte gradually increases to a value in the range 1.26–1.28.

The instrument used to test the 'condition' of the charge or discharge of a cell of the battery is known as a **hydrometer**, and consists of a glass tube containing a calibrated float. When the hydrometer is partly-filled with electrolyte, the level of the calibrated float indicates the density of the electrolyte. Typical values of density are:

Condition	Density
fully charged	1.26 to 1.28 (indicated as 1260 to 1280)
half charged	1.25 (indicated as 1250)
discharged	1.15–1.2 (indicated as 1150 to 1200)

At the end of the charging process (described below), each cell in the battery has an e.m.f. of about 2.5 V. When the battery has had a period of rest and the electrolyte has diffused throughout each cell, the e.m.f. per cell falls to about 2–2.1 V.

During the early part of the discharge period (say the first hour), the terminal voltage of the battery falls fairly quickly to give an equivalent voltage per cell of about 2 V. During the major part of the discharge period the terminal voltage falls fairly slowly because the electrolyte in the pores of the separators is used up more quickly than fresh electrolyte can flow in. At the end of the discharge period the terminal voltage falls to about 1.8 V per cell, by which time the plates are covered with lead sulphate ($PbSO_4$) and the specific gravity has fallen to 1.2 or lower.

During the charging process, the lead sulphate in the paltes is converted into lead peroxide (in the positive plates) and lead (in the negative plates), and the specific gravity gradually increases (indicating the formation of more sulphur acid). The charging process is complete when the lead sulphate is exhausted, and is indicated by a sharp rise in the terminal voltage to give an e.m.f. per cell of about 2.5 V. At this time electrolysis of the water in the electrolyte occurs, causing hydrogen and oxygen gas to be produced at the negative and positive plates, respectively (see section 2.4 for details). When this occurs in a cell, it is said to be **gassing**.

The nickel–iron (NiFe) alkaline accumulator

This is known in the US as the **Edison** accumulator. This type of battery has a positive plate made of nickel oxide, a negative plate made of iron, the electrolyte being a dilute solution of potassium hydroxide with a small amount of lithium hydroxide.

This type is more robust than the lead–acid accumulator but produces a lower e.m.f. per cell (about 1.2 V). The efficiency of the charge–discharge process is lower than that of the lead–acid accumulator, being about 60 per cent compared with about 75 per cent for the lead-acid type.

The Nickel–Cadmium (NiCd) alkaline accumulator

The electrodes and the electrolyte in this accumulator are generally similar to those in the nickel–iron accumulator with the exception that the negative electrode is either made of cadmium or a mixture of iron and cadmium. The average e.m.f. per cell is about 1.2 V; this type of accumulator is widely used in radios, televisions, power tools, etc. Nickel–cadmium batteries are used as 'on-board' power supplies in many computer systems; their function being to maintain a power supply to electronic memories in the event of a power failure.

An interesting feature of the NiCd cell is that the chemical expression, KOH, for the electrolyte does not appear in the charge–discharge equation. This means that the specific gravity of the electrolyte does not change with the state of the charge of the cell; that is to say, the specific gravity of the electrolyte does not give an indication of the state of charge.

Other storage batteries

Research not only into specialised electronic applications but also into electric traction (the 'electric' car) has given rise to a number of new accumulators. These include the **zinc-chloride cell** (about 2.1 V per cell), the **lithium–iron sulphide cell** (about 1.6 V per cell) and the **sodium-sulphur cell**. Another interesting cell is the **plastic cell** which uses a conductive polymer; this cell is claimed to have a much reduced weight and volume combined with a higher capacity than the lead–acid cell.

Capacity of an accumulator

The capacity of an accumulator is expressed in terms of the number of *ampere–hours (Ah) that may be taken from it* during the discharge period. Thus, a 20 Ah battery is capable of supplying a current of 2 A for a period of time given by

$$\text{time} = \frac{\text{ampere–hour capacity (Ah)}}{\text{discharge current (A)}}$$

$$= \frac{20}{2} = 10\,\text{h}$$

In addition to the ampere–hour rating, the manufacture specifies the maximum discharge current. If this current is exceeded, the ampere–hour rating of the accumulator is reduced. For example, in the example shown the capacity of the battery may be reduced to, say, 10 Ah if the maximum current is stated as 2 A and the current drawn is 4 A.

The charge–discharge ampere–hour efficiency of the battery must also be taken into account when determining the charge given to a battery as follows:

$$\text{ampere–hour charge} = \frac{\text{ampere–hour discharge}}{\text{efficiency}}$$

For a lead-acid battery whose capacity is 20 Ah and whose charge-discharge efficiency is 75 per cent, the amount of charge needed, in ampere–hours is

$$20 \div \frac{75}{100} = 20 \times \frac{100}{75} = 26.7 \text{ Ah}$$

2.14 THERMOELECTRICITY

There are a number of ways in which thermal effects can directly produce electricity and we shall consider two of them.

The Seebeck effect – the thermocouple

When two different metals are brought into contact with one another, it is found that electrons can leave one of the metals more easily than they can leave the other metal. This is because of the difference in what is known as the **work function** of the two metals. Since electrons leave one metal and are gained by the other, a potential difference exists between the two metals; this e.m.f. is known as the **contact potential** or **contact e.m.f.**

If two metals, say copper and iron, are joined at two points as shown in Figure 2.7, and both junctions are at the same temperature, the contact potentials cancel each other out and no current flows in the loop

fig 2.7 *the Seebeck effect*

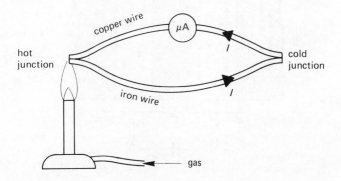

of wire. However, Thomas Johann Seebeck (1770-1831) discovered that if the two junctions are kept at different temperatures, there is a drift of electrons around the circuit, that is to say, current flows.

The magnitude of the voltage produced by this method is small – only a few millivolts per centigrade degree – but it is sufficient to be able to measure it. The current flow is a measure of the temperature of the 'hot' junction (the 'cold' junction meanwhile being maintained at a low temperature – often 0 °C). Each junction is known as a **thermocouple**, and if a number of thermocouples are connected in series so that alternate junctions are 'hot' and the other junctions are 'cold', the total e.m.f. is increased; this arrangement is known as a **thermopile**.

The Peltier effect

This was discovered by Pierre Joseph Peltier (1788-1842), who showed that when an electric current flows across the junction of two different substances, heat is either absorbed or liberated at the junction. The 'direction' of heat flow is the same as the flow of the 'majority' charge carriers in the material; these charge carriers may be either electrons or holes (depending on the material).

The basis of one such device is shown in Figure 2.8. It comprises two surfaces connected in one case by an *n*-type semiconductor and in another

fig 2.8   *the basis of a thermoelectric heat-transfer device*

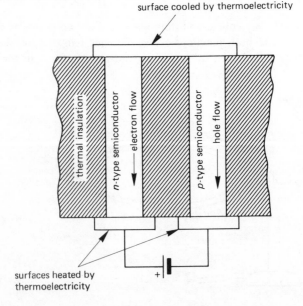

case by a *p*-type semiconductor. In the *n*-type material, the 'majority' charge carriers are electrons, and in the *p*-type material they are holes.

When current flows in the circuit, the current in the *n*-type semiconductor is largely electron flow (towards the positive pole of the battery), so that heat is transferred from the upper surface to the lower surface via the *n*-type material. At the same time, the current in the *p*-type semiconductor is largely hole flow (towards the negative pole of the battery), so that heat is once more transferred from the upper surface to the lower surface by the *p*-type material. In this way, the flow of electricity cools the upper surface while heating the lower surfaces.

2.15 THE HALL EFFECT

When experimenting in 1879 with current flowing in a strip of metal, E. M. Hall discovered that some of the charge carriers were deflected to one of the faces of the conductor when a strong magnetic field was applied. This gave rise to an e.m.f. (the **Hall voltage**) between opposite faces of the conductor. The e.m.f. is only a few microvolts in the case of a metal conductor, but is much larger when the current flows in a semiconductor.

The general principle is illustrated in Figure 2.9. As mentioned earlier, current flow in a *p*-type semiconductor is largely due to the movement of holes or mobile positive charge carriers. When a magnetic flux passes through the semiconductor in the direction shown in Figure 2.9, the holes

fig 2.9 *Hall effect in a p-type semiconductor*

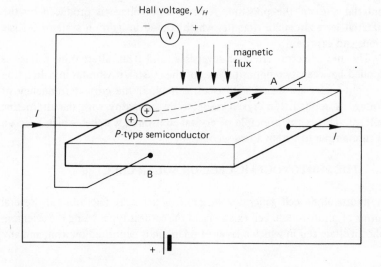

experience a force which deflects them towards face A of the semiconductor (the reader should refer to Fleming's Left-Hand Rule in Chapter 9 for details on how to determine the direction of the force). This gives rise to the Hall voltage, V_H, which makes face A positive with respect to face B.

If the current flow through the semiconductor is maintained at a constant value, the *Hall voltage is proportional to the magnetic flux density passing through the semiconductor.*

One application of the Hall effect is the measurement of magnetic fields; instruments used for this purpose are known as **magnetometers**. Another application is to **contactless keys** on a computer keyboard (or any keyboard for that matter); the semiconductor element is fixed to the frame of the keyboard and the magnet is mounted in the movable keypad. When the key is pressed, the magnet is moved down to the semiconductor, the resulting Hall voltage being amplified and transmitted to the computer.

2.16 THE PIEZOELECTRIC EFFECT

Certain crystals and semiconductors produce an e.m.f. between two opposite faces when the mechanical pressure on them is either increased or reduced (the polarity of the e.m.f. is reversed when the pressure changes from an increase to a decrease). This e.m.f. is known as the **piezoelectric e.m.f.**

This effect is used in a number of devices including **semiconductor strain gauges** and **crystal pick-ups** for gramophones. As the mechanical pressure on the crystal is altered, a varying voltage which is related to the pressure is produced by the crystal. The voltage can be as small as a fraction of a volt or as large as several thousand volts depending on the crystal material and on the pressure. A very high voltage is produced by the material lead zirconate titanate, which is used in ignition systems for gas ovens and gas fires.

The piezoelectric effect is reversible, and if an alternating voltage is applied between two opposite faces of the crystal, it vibrates in a direction at right-angles to the applied electric field. If the correct frequency of voltage is applied, the crystal **resonates** in sympathy with the alternating voltage. This is the principle of **crystal-controlled** watches which keep an accurate time to within a few seconds a year.

2.17 THE PHOTOVOLTAIC CELL OR SOLAR CELL

A **photovoltaic cell** generates an e.m.f. when light falls onto it. Several forms of photovoltaic cell exist, one of the earliest types being the selenium photovoltaic cell in which a layer of selenium is deposited on iron, and any light falling on the selenium produces an e.m.f. between the selenium and the iron.

Modern theory shows that the junction at the interface between the two forms what is known as a semiconductor p–n junction in which one of the materials is p-type and the other is n-type. The most efficient photovoltaic cells incorporate semiconductor p–n junctions in which one of the regions is a very thin layer (about 1 μm thick) through which light can pass without significant loss of energy. When the light reaches the junction of the two regions it causes electrons and holes to be released, to give the electrovoltaic potential between the two regions.

Applications of the photovoltaic cell or solar cell include camera exposure-meters, electricity supply to batteryless calculators and to satellites.

SELF-TEST QUESTIONS

1. What is meant by the electrochemical effect? Write down a list of applications of this effect.
2. Explain what is meant by a 'positive ion' and a 'negative ion', and give an example showing where they occur in electrolysis.
3. State Faraday's Laws of electrolysis. Explain how the laws are used to calculate the mass of material deposited in an electrochemical process.
4. Explain the difference (i) between a 'cell' and a 'battery', (ii) between a primary cell and a secondary cell.
5. A battery has a no-load terminal voltage of 12 V. When a current of 20 A is drawn from the battery, the terminal voltage falls to 10 V. Determine the internal resistance of the battery.
6. Describe the operation of (i) one type of dry cell and (ii) one type of storage battery.
7. Explain what is meant by (i) the Seebeck effect, (ii) the Peltier effect, (iii) the Hall effect, (iv) the piezoelectric effect and (v) the photovoltaic effect.

SUMMARY OF IMPORTANT FACTS

Pure liquids are good insulators, but liquids **containing salts** conduct electricity.

An **ion** is an atom which has either *lost an electron* (a **positive ion**) or has *gained an electron* (a **negative ion**).

Electrolysis is the process of decomposing an electrolyte by the passage of electric current through it; this results in chemical action at the **electrodes**, that is, the **anode** and the **cathode**. Electrolysis is the basis not only of many forms of chemical **extraction** and **refining** but also of the **electro-plating** industry.

A **catalyst** is a material which, when present in association with certain other materials, causes a chemical action to occur but is not itself affected.

The laws which govern electrolysis are described by **Faraday's laws**.

An electrical **cell** consists of two sets of **plates** immersed in an **electrolyte**. Cells can be either **dry** or **wet**. A **primary cell** cannot be recharged, but a **secondary cell** can be recharged. A **battery** is an interconnected group of cells. All cells have an **internal resistance** whose value is reduced by the use of a **depolariser**.

Electricity can be produced by a number of different methods including **chemical action**, **thermoelectricity**, the **Hall effect**, the **piezoelectric effect** and the **photovoltaic effect**.

RESISTORS AND ELECTRICAL CIRCUITS

3.1 RESISTOR TYPES

A resistor is an element whose primary function is to limit the flow of electrical current in a circuit. A resistor is manufactured either in the form of a **fixed resistor** or a **variable resistor**, the resistance of the latter being alterable either manually or electrically. Many methods are employed for the construction of both fixed and variable resistors, the more important types being described in this chapter.

3.2 FIXED RESISTORS

Carbon composition resistors or carbon resistors
The resistive element of this type is manufactured from a mixture of finely ground carbon compound and a non-conducting material such as a ceramic powder. The material is moulded into the required shape (cylindrical in Figure 3.1) and each end is sprayed with metal to which a wire is connected (alternatively, metal end-caps are pressed onto the resistor).

fig 3.1 *a carbon composition resistor*

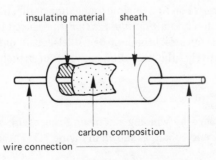

fig 3.2 *a film resistor*

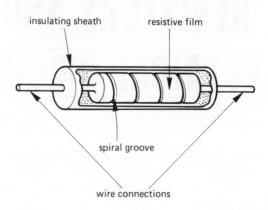

insulating sheath   resistive film

spiral groove

wire connections

Film resistors

The resistive element is a thin film of resistive material which is deposited on an insulating 'former (see Figure 3.2). The resistance of the resistor is increased to the desired value at the manufacturing stage by cutting a helical groove in the film (thereby increasing the length of the resistive path). There are three popular types of film resistor, namely *carbon film* (or *cracked carbon resistors*), *metal oxide film resistors* and *metal film resistors*.

A *carbon film resistor* is formed by depositing a thin film of carbon on an insulating rod. This type of resistor is subject to damage by atmospheric pollution, and is protected by several layers of lacquer or plastic film. *Metal oxide film resistors* (also known as *oxide film resistors*) comprise a tin-oxide film deposited on a ceramic former; this type of resistor can be run at a higher temperature than can a carbon film resistor. A *metal film resistor* is formed by evaporating a nickel–chromium alloy onto a ceramic substrate.

A form of resistor known as a *thick film resistor* or *cermet resistor* is manufacturted by depositing a 'thick' film (about 100 times thicker than on a carbon film resistor) of a mixture of a **cer**amic and **met**al (which is abbreviated to **cermet**) on to a ceramic foundation or substrate. Yet another type of resistor known as a *thin film resistor* is produced by evaporating a thin film of either nickel–chromium or nickel–cobalt onto an insulating substrate; alternatively, the thin film may be of tantalum doped with aluminium.

Wirewound resistors

This type is produced by winding a ceramic former with wire made from an alloy of nickel and cobalt, as shown in Figure 3.3. These materials have

fig 3.3 *a wirewound resistor*

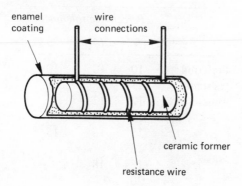

good long-term stability, and their resistance is reasonably constant despite temperature variation.

3.3 PREFERRED VALUES OF RESISTANCE FOR FIXED RESISTORS

The accuracy to which a resistor is manufactured is, to some extent, related to the cost of the resistor; the greater the accuracy, the greater the cost. Within the limits of manufacturing cost, every resistor is produced to within a certain percentage tolerance; typical tolerances are 5 per cent, 10 per cent and 20 per cent. Thus a nominal 10 Ω resistor with a 5 per cent tolerance will have a value in the range 9.5–10.5 Ω; if the tolerance was 20 per cent, its value would lie in the range 8.0–12.0 Ω.

The international range of 'nominal' preferred values of resistance used in electrical and electronic engineering is given in Table 3.1. The values are selected so that, within each tolerance range, the resistance of a resistor at its lower tolerance limit is approximately equal to the value of the resistance of the next lower resistor at its upper tolerance limit. Similarly, its resistance at its upper tolerance limit is approximately equal to the value of the next higher resistance at its lower tolerance limit.

Table 3.1 uses a starting value of 10 Ω, but it could equally well be, say, 0.1 Ω, or 1.0 Ω, or 1.0 kΩ or 1.0 MΩ, etc.

3.4 RESISTANCE COLOUR CODE

The value of the resistance of many resistors (but not wirewound types) used in low-power electrical circuits is indicated on the resistor itself by means of a **colour band** coding. This is an international coding, and is given in Table 3.2, and illustrated in Figure 3.4.

Table 3.1 *Preferred values of resistors*

	Percentage tolerance		
	5	*10*	*20*
Preferred	10	10	10
resistance	11		
value	12	12	
	13		
	15	15	15
	16		
	18	18	
	20		
	22	22	22
	24		
	27	27	
	30		
	33	33	33
	36		
	39	39	
	43		
	47	47	47
	51		
	56	56	
	62		
	68	68	68
	75		
	82	82	
	91		

fig 3.4 *resistor colour-band coding*

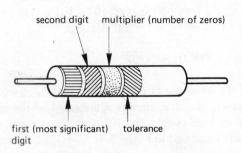

second digit   multiplier (number of zeros)

first (most significant) digit   tolerance

Table 3.2 *Resistance colour code*

Colour	Value of first and second digits	Multiplier	Number of zeros to left of decimal point	Tolerance (per cent)
No tolerance band				20
Silver		0.01		10
Gold		0.10		5
Black	0	1.00		
Brown	1	10.00	1	1
Red	2	10^2	2	2
Orange	3	10^3	3	3
Yellow	4	10^4	4	4
Green	5	10^5	5	
Blue	6	10^6	6	
Violet	7	10^7	7	
Grey	8	10^8	8	
White	9	10^9	9	

The reader will find the following mnemonic useful as an aid to remembering the colour sequence. It is based on the old British Great Western Railway Company, the first letter in each word in the mnemonic is the first letter of the colour used to identify the value of the first and second digits of the resistor value.

Bye Bye Rosie, Off You Go, Bristol Via Great Western

corresponding to the colours

Black Brown Red Orange Yellow Green Blue Violet Grey White

The use of the colour code is illustrated in the example given in Table 3.3. This resistor is a 4700 Ω resistor with a tolerance of 10 per cent in other words, its value lies in the range 4230 Ω–5170 Ω.

Table 3.3 *Example of colour coding*

Band	Colour	Comment
First	Yellow	Most significant digit = 4
Second	Violet	Least significant digit = 7
Third	Red	Multiplier = 2 (two zeros)
Fourth	Silver	Tolerance = 10 per cent

3.5 VARIABLE RESISTORS, RHEOSTATS AND POTENTIOMETERS

A **variable resistor** or **rheostat** is a resistor whose value can be varied (see Figure 3.5(a)); it is sometimes connected in the form of a **potentiometer** (abbreviated to '**pot**') – see Figure 3.5(b) – which can be used as a *voltage divider*.

A variable resistor consists of a resistive element which either has several tapping points on it or has a sliding contact (known as the **slider** or **wiper**). The unused connection can either be left unconnected or it may be connected to the tapping point or wiper (see Figure 3.5(a)). When used as a potentiometer, it is used to supply an electrical load with voltage V_2, the output voltage being given by the equation

$$V_2 = \frac{R_2 V_S}{R} \text{ volts (V)}$$

where V_S is the supply voltage.

In a **linear potentiometer**, the resistance between the wiper and the 'bottom' end of the pot increases in a linear or 'straight line' manner as the slider is moved along the resistive element. Thus, doubling the movement of the slider, doubles the resistance between the slider and the bottom end of the pot.

Many potentiometers used in electronic circuits have a resistance whose value varies in a *logarithmic* ratio with the movement of the slider (typical applications of this type include volume controls on radios and TVs). Such potentiometers are known as **logarithmic potentiometers** or **log pots**. This type of pot allows the response of the human ear to be matched to the electronic equipment.

fig 3.5   *(a) variable resistor, (b) a resistive potentiometer*

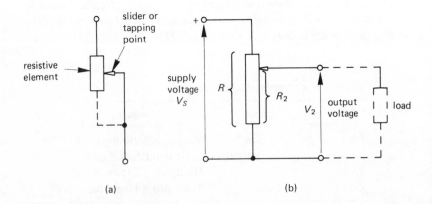

(a)   (b)

The name used to describe variable resistors is related to the track shape, the most popular types being

(a)   rectilinear
(b)   arc
(c)   helical or multi-turn.

A **rectilinear variable resistor** has a slider which can be moved in a straight line along the resistor (see Figure 3.6(a). A **single-turn** or **arc shaped variable resistor** has its resistive element in the form of an arc (see Figure 3.6(b)), the arc angle being on the range 300° to 330°. A few highly specialised potentiometers provide 360° rotation for such purposes as trigonometric function generators. A **helical track** or **multi-turn potentiometer** (see Figure 3.6(c)) has its resistive element in the form of a multi-turn helix. A ten-turn helical track potentiometer gives an equivalent angular rotation of 3600°.

Types of resistive element used in fixed and variable resistors
The main types in use are

(a)   wirewound
(b)   carbon
(c)   cermet
(d)   conductive plastic

Types (a), (b) and (c) have been described earlier in connection with fixed resistors. A *conductive plastic* resistive track consists of carbon particles distributed throughout a thermosetting resin; the resulting track provides a life expectancy which exceeds that of all other types. Unfortunately, the contact resistance between the slider and the track is fairly high, and this limits the output current which can be taken from the slider.

3.6 RESISTANCE OF A CONDUCTOR

The resistance of an electrical circuit depends on several factors, including the dimensions of the conductor, that is, its length and its cross-sectional area.

Consider for a moment a conductor having a length l and area a whose resistance is R ohms (see Figure 3.7(a)). If two such conductors are connected in series with one another (see Figure 3.7(b)) to give an effective length of $2l$ then, in order to force the same value of current, I, through the circuit, double the potential drop is required across the two series-connected conductors. That is to say, **doubling the length of the con-**

42

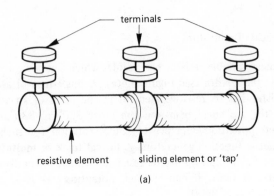

(a)

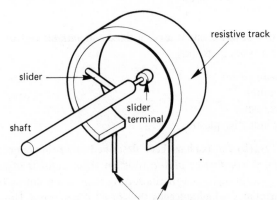

(b)

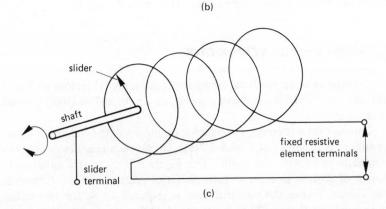

(c)

fig 3.7 *(a) original conductor having resistance R; effect of (b) an increase in length; and (c) an increase in area of the resistance of a conductor*

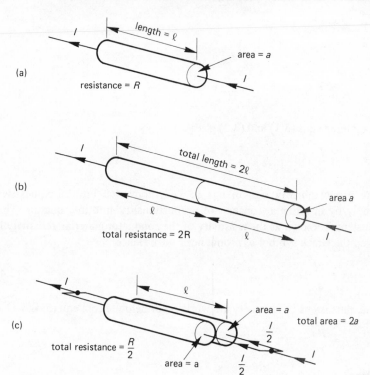

ductor has doubled its resistance. Hence, **the resistance is proportional to the lenngth of the conductor,** or

resistance α length of the conductor

or

$$R \propto l \qquad (3.1)$$

If the two identical conductors of resistance R are connected in parallel with one another as shown in Figure 3.7(c), each conductor carries one-half of the total current, that is, $\frac{I}{2}$. Since this current flows through a resistor of resistance R, the p.d. across each conductor is therefore one-half that across the conductor in diagram (a) [which carries current I]. Since the total current carried by the parallel combination is I, and the p.d. across the combination is reduced to one-half the value across the conductor in diagram (a), the resistance of the parallel combination in Figure 3.7(c) is one-half that of the conductor in diagram (a). This arises because *the total area of the conducting path has doubled*; that is, **the resistance is inversely proportional to the area, *a*, of the conductor**, or

resistance $\alpha \; \dfrac{1}{\text{area}}$

or

$$R \; \alpha \; \frac{1}{a} \qquad (3.2)$$

Combining eqns (3.1) and (3.2) gives

$$R \; \alpha \; \frac{l}{a} \qquad (3.3)$$

The proportionality sign in eqn (3.3) is converted into an equals sign simply by inserting a 'constant of proportionality' into the equation. This constant is known as the **resistivity** of the conductor material; resistivity is given the Greek symbol ρ (pronounced rho). Hence

$$R = \frac{\rho l}{a} \; \Omega \qquad (3.4)$$

The dimensions of resistivity are worked out below. From eqn (3.4)

$$\rho = \frac{Ra}{l}$$

hence

dimensions of ρ = dimensions of $\dfrac{Ra}{l}$

$$= \frac{\text{no. of ohms} \times \text{area}}{\text{length}}$$

$$= \frac{\text{no. of ohms} \times [\text{length} \times \text{length}]}{\text{length}}$$

$$= \text{no. of ohms} \times \text{length or ohm metres } (\Omega \, \text{m})$$

Typical values for the resistivity of materials used in electrical engineering are given in Table 3.4. The materials manganin and constantan have high resistivity combined with a low temperature coefficient of resistance (see section 3.8), and are widely used in ammeter shunts and in voltmeter multiplier resistors (see also Chapter 4). Nichrome has a very high resistivity and is used as a conductor in heating elements in fires and ovens, and operates satisfactorily at temperatures up to 1100 °C. The reader should

Table 3.4 *Resistivity of metals at 0°C*

Material	Resistivity (Ω m)
Silver	1.47×10^{-8}
Copper	1.55×10^{-8}
Aluminium	2.5×10^{-8}
Zinc	5.5×10^{-8}
Nickel	6.2×10^{-8}
Iron	8.9×10^{-8}
Manganin	41.5×10^{-8}
Constantan	49.0×10^{-8}
Nichrome	108.3×10^{-8}

note that the resistivities listed in Table 3.4 may vary with temperature. For example, the resistivity of copper increases to 1.73×10^{-8} Ωm at 20 °C, but the resistivity of constantan remains unchanged when the temperature changes from 0° to 20 °C.

Insulators have much higher values of resistivity, typical values in Ωm being

Glass	10^9-10^{12}
Mica and mineral oil	10^{11}-10^{15}
Plastic	10^7-10^9
Wood	10^8-10^{11}
Water (distilled)	10^2-10^5

Example

Determine the resistance of a 200 m length of copper wire of diameter 1 mm at a temperature of 20 °C. The resistivity of copper at this temperature is

1.73×10^{-8} Ωm.

Solution

$l = 200$ m, $\rho = 1.73 \times 10^{-8}$ Ωm, $d = 1$ mm $= 10^{-3}$ m

Note

When dealing with submultiples such as mm (and with multiples for that matter), the dimensions should be converted into the basic unit (the metre in this case) before using the value in any equation. This avoids a possible cause for error later in the calculation.

$$\text{Area of conductor} = \frac{\pi \times d^2}{4} = \frac{\pi \times (10^{-3})^2}{4}$$

$$= 0.785 \times 10^{-6} \text{ m}^2$$

hence

$$\text{Resistance, } R = \frac{\rho l}{a}$$

$$= 1.73 \times 10^{-8} \times \frac{200}{0.785} \times 10^{-6} \text{ (m)}$$

$$= 4.41 \text{ } \Omega \text{ (Ans.)}$$

3.7 CONDUCTIVITY AND CONDUCTANCE

The **conductivity** of a material is the **reciprocal of its resistivity**. It is given the Greek symbol σ (sigma) and has the units siemens per metre (S/m). Thus, at $0\,^{\circ}$C, copper has a conductivity of

$$\sigma = \frac{1}{\rho} = \frac{1}{1.55 \times 10^{-8}}$$

$$= 64.52 \times 10^6 \text{ S/m}$$

Also, the **conductance**, G, of a material is the **reciprocal of its resistance** and is

$$G = \frac{1}{R} = \frac{1}{\rho\, l/a} = \frac{1}{\rho} \times \frac{a}{l} = \sigma \times \frac{a}{l}$$

3.8 EFFECT OF TEMPERATURE CHANGE ON CONDUCTOR RESISTANCE

As the temperature of a conductor rises, the atomic nuclei gain energy and become 'excited'. When this happens, the current in the circuit (which can be regarded as electrons in motion) experience increasing difficulty in moving through the conductor. That is, **increase in temperature causes an increase in conductor resistance**. Conversely, *a reduction in temperature causes a decrease in the resistance of the conductor*. The change in resistance is proportional to the change in temperature over the normal operating range of temperature.

This effect is shown graphically in Figure 3.8; the resistance of the conductor represented by the graph at $0\,^{\circ}$C is given the value R_0, at temperature $\theta_1\,^{\circ}$C the resistance is R_1, at $\theta_2\,^{\circ}$C it is R_2, etc.

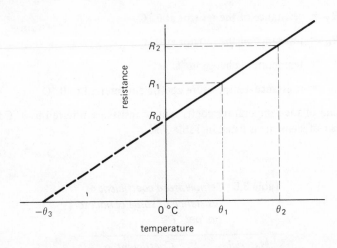

fig 3.8 *change in the resistance of a conductor with change in temperature*

The change in resistance of the resistor as the temperature changes from, say, $0\,^\circ$C to temperature θ_1, expressed as a fraction of its original resistance R_0 is called the **temperature coefficient of resistance** referred to the original temperature ($0\,^\circ$C in this case). This coefficient is given the Greek symbol α (alpha), where

$$\alpha_0 = \frac{R_1 - R_0}{(\theta_1 - 0)R_0} = \frac{R_1 - R_0}{\theta_1 R_0} \tag{3.5}$$

Note: the value of α is given the value α_0 in the above equation since the original temperature was $0\,^\circ$C.

The units of α can be determined from the above equation as follows:

$$\text{Units of } \alpha = \frac{\text{resistance}}{(\text{temperature} \times \text{resistance})}$$

$$= \frac{1}{\text{units of temperature}}$$

That is, α is expressed in 'per degree C' or $(^\circ\text{C})^{-1}$.
From eqn (3.5)

$$R_1 - R_0 = \alpha_0 \theta_1 R_0$$

or

$$R_1 = R_0 + \alpha_0 \theta_1 R_0 = R_0(1 + \alpha_0 \theta_1) \tag{3.6}$$

In general, eqn (3.6) can be rewritten in the form

$$R_T = R_0(1 + \alpha_0 \theta) \tag{3.7}$$

where

R_T = resistance of the resistor at θ °C

R_0 = resistance of the resistor at 0 °C

θ = temperature change in °C

α_0 = resistance–temperature coefficient referred to 0 °C

The value of the temperature coefficient of resistance referred to 0 °C for a number of elements is listed in Table 3.5.

Table 3.5 *Temperature coefficient of resistance of some conductors (in 'per °C')*

Material	Coefficient, α
Aluminium	0.00435
Copper	0.00427
Iron	0.00626

Example

The resistance of a coil of copper wire at 20 °C is 25 Ω. What current will it draw from a 10-V supply when in a room at 0 °C? The temperature coefficient of resistance of the copper being 0.00427 per °C referred to 0 °C.

Solution

$R_T = 25\ \Omega, \theta_1 = 20$ °C, $E = 10$ V, $\alpha_0 = 0.00427$ (°C)$^{-1}$

From eqn (3.7)

$$R_0 = \frac{R_T}{(1 + \alpha_0\theta)} = \frac{25}{(1 + [0.00427 \times 20])}$$

$$= 23.033\ \Omega$$

Hence

$$\text{current at 0 °C}, I_0 = \frac{E}{R_0} = \frac{10}{23.033}$$

$$= 0.434\ \text{A (Ans.)}$$

3.9 SUPERCONDUCTIVITY

It is shown in Figure 3.8 that if the temperature of a conductor is progressively reduced, its resistance would reach zero at some temperature $-\theta_3$. However, scientific tests have shown that, for a limited range of conductors, the resistance falls to a value which is too small to be measured at a temperature within $10°$ of absolute zero temperature $(-273°C)$ – see Figure 3.9. Included in these are tin (within $3.7°$ of absolute zero), mercury (within $4.1°$) and lead (within $7.2°$).

3.10 TEMPERATURE EFFECTS ON INSULATORS AND ON SEMICONDUCTORS

Insulators and semiconductors behave in a different way when the temperature increases, because their resistivity *decreases*. That is: **the resistance of an insulator and of a semiconductor decreases with temperature increase**, (their resistance–temperature coefficient is negative!). This feature can be used to advantage as the following example shows.

One example of this effect occurs in a **thermistor**, which is a THERMally sensitive resISTOR whose resistance alters with temperature; a **negative temperature coefficient (n.t.c.) thermistor** is one whose resistance reduces with increase in temperature. A thermistor is used in the cooling-water temperature-measuring circuit of a car or lorry; it is inserted in the cooling water and connected in series with the battery and temperature gauge. As the water temperature rises, the resistance of the n.t.c. thermistor falls and

fig 3.9 *superconductivity*

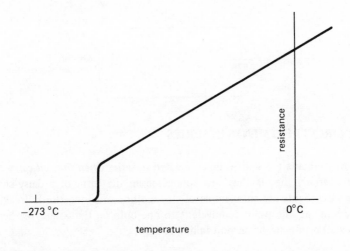

allows more current to flow through the temperature gauge; this causes the gauge to indicate variations in water temperature.

3.11 WHAT IS AN ELECTRICAL CIRCUIT?

The function of an electrical circuit, no matter how complex, is to provide a path for the flow of electrical current. In its simplest form, it consists of a battery which is connected via a pair of wires to a load (a heater in the example shown in Figure 3.10(a)). The **circuit diagram** in Figure 3.10(b) represents the physical circuit in diagram (a), the heating element being shown as a resistor (which is the rectangular symbol) in Figure 3.10(b). You will note that the **current**, I, flows *out of the positive pole* of the battery and *returns to the negative pole*; this notation is adopted in both electrical and electronic circuits.

fig 3.10 *a simple electrical circuit*

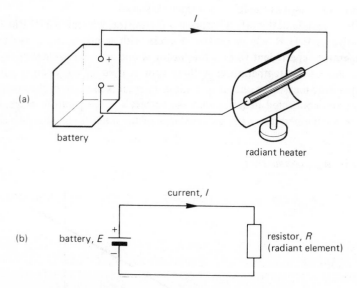

(a)

battery

radiant heater

current, I

(b) battery, E resistor, R
(radiant element)

3.12 CIRCUIT ELEMENTS IN SERIES

Circuit elements are said to be connected in **series** *when they all carry the same current*, that is they are connected in the form of a daisy-chain (Xmas-tree lights are an example of this method of connection). Breaking the circuit at one point (equivalent to one bulb on the Xmas tree being removed) results in the current falling to zero.

The *series connection of three resistors* is shown in diagram (a) of Figure 3.11. Since the current must pass through each resistance in the circuit, the **effective resistance** or **total resistance**, R_S, of the series circuit is the sum of the individual resistors in the circuit. That is

$$R_S = R_1 + R_2 + R_3$$

In a series circuit, the total resistance is always greater than the largest individual resistor in the circuit

For Figure 3.11(a), the effective resistance is

$$R_S = 10 + 5 + 2 = 17 \ \Omega$$

In the general case where 'n' resistors are connected in series with one another, the total resistance of the circuit is

$$R_S = R_1 + R_2 + R_3 + \ldots + R_n$$

When three *e.m.fs are connected in series* with one another (see Figure 3.11(b)), some care must be taken when determining both the magnitude and polarity of the resulting voltage. For example, to determine the potential of point B with respect to point A (which is written as V_{BA}), the procedure is as follows. First, it is necessary to indicate the polarity of each e.m.f. on the diagram *by means of a 'potential' arrow drawn by the side of each e.m.f.* The method adopted here is to draw an arrow *pointing towards the positive pole of the battery*, that is, pointing in the direction in which the e.m.f. would urge current to flow. Second, the starting at point A we *add together* the e.m.fs whose *potential arrows point in the direction in which we are moving*, that is, arrows pointing from A towards

fig 3.11 *series connection of (a) resistors, (b) e.m.fs*

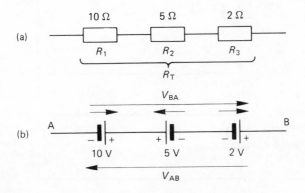

B and we *subtract* the em.fs whose *potential arrows point in the opposite direction.* Thus

$$V_{BA} = +10 - 5 + 2 = 7 \text{ V}$$

Some circuits may be complicated by the fact that the battery supplying the circuit may itself have some internal resistance (see also Chapter 2). This resistance may significantly affect the current in the circuit, and the following is one method of dealing with this problem.

Suppose that the battery in Figure 3.10(a) has some internal resistance; the circuit diagram is redrawn in Figure 3.12(a) to show the internal resistance r as an integral part of the battery. The circuit can be drawn as shown in Figure 3.12(b), which combines the load resistance with the internal resistance of the battery to give a total circuit resistance of $(R + r)$ ohms. The current in the circuit is calculated from the equation

$$I = \frac{\text{e.m.f. } (E)}{\text{total resistance } (R + r)}$$

Suppose that $R = 10 \ \Omega$ and that either of two 10-V batteries may be connected to R. Calculate the current in both cases if the internal resistance of one battery is 5 Ω and of the other is 0.005 Ω.

fig 3.12 *a circuit which account for the internal resistance, r, of the battery*

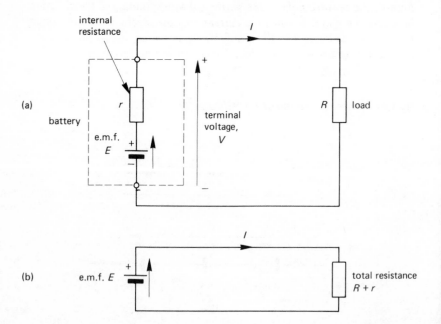

(a) For r = 5 Ω

Total resistance = $R + r = 10 + 5 = 15$ Ω

$$\text{Current} = \frac{E}{(R + r)} = \frac{10}{15} = 0.667 \text{ A (Ans.)}$$

(b) For r = 0.005 Ω

Total resistance = $R + r = 10 + 0.005 = 10.005$ Ω

$$\text{Current} = \frac{E}{(R + r)} = \frac{10}{10.005} = 0.9995 \text{ A (Ans.)}$$

Clearly, a battery with a large internal resistance gives a low current. You will have experienced the case of, say, a car or torch battery which is 'flat'; its internal resistance is high and it can supply little energy to the connected load.

3.13 RESISTORS IN PARALLEL

Resistors are said to be connected in **parallel** with one another *when they each have the same voltage across them.* Electric lights in a house are all connected in parallel to the mains electricity supply; when one lamp is switched off, it does not affect the supply voltage or current to the other lamps.

Figure 3.13 shows a number of resistors connected in parallel with one another between the points A and B. The **effective resistance**, R_P, is calculated from the equation

$$\frac{1}{R_P} = \frac{1}{R_1} + \frac{1}{R_2} + \frac{1}{R_3} + \ldots + \frac{1}{R_n}$$

If, for example the three resistors $R_1 = 10$ Ω, $R_2 = 5$ Ω and $R_3 = 2$ Ω are connected in parallel with one another, then

fig 3.13 *parallel-connected resistors*

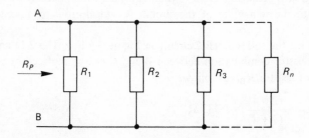

$$\frac{1}{R_P} = \frac{1}{10} + \frac{1}{5} + \frac{1}{2} = 0.8$$

hence

$$R_P = \frac{1}{0.8} = 1.25 \ \Omega$$

In a parallel circuit, the total resistance is always less than the smallest individual resistance in the circuit

In the *special case* of two resistors R_1 and R_2 connected in parallel, the effective value of the parallel circuit is given by

$$R_P = \frac{R_1 R_2}{R_1 + R_2}$$

Example

Calculate the effective resistance of a parallel circuit containing a 50-ohm and a 20-ohm resistor.

Solution

Given that $R_1 = 50 \ \Omega, R_2 = 20 \ \Omega$

$$R_P = \frac{(R_1 R_2)}{(R_1 + R_2)}$$

$$= \frac{(50 \times 20)}{(50 + 20)} = \frac{1000}{70} = 14.29 \ \Omega \ (\text{Ans.})$$

Note: the value of R_P is less than either R_1 or R_2.

3.14 SERIES-PARALLEL CIRCUITS

Many practical circuits contain both series and parallel combinations of resistors. Each circuit must be treated on its merits, appropriate series and parallel groups being converted to their equivalent values before combining them with the remainder of the circuit. A typical problem is considered below.

Consider the series-parallel circuit in Figure 3.14(a). The 2 Ω and 4 Ω parallel-resistor combination between B and C can be replaced by a single equivalent resistor R_{BC} as follows

$$R_{BC} = \frac{(2 \times 4)}{(2 + 4)} = 1.333 \ \Omega$$

fig 3.14 *a series-parallel circuit*

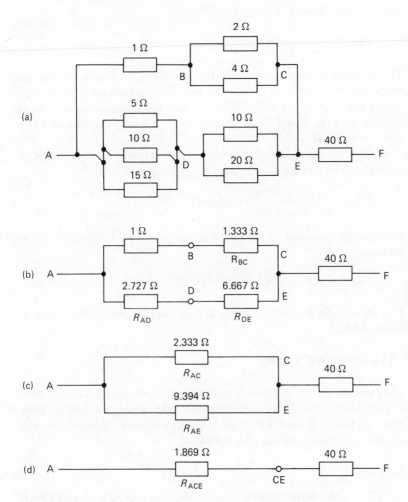

The three parallel-connected resistors in the bottom branch between A and D can be replaced by the single resistor R_{AD} calculated below

$$\frac{1}{R_{AD}} = \frac{1}{5} + \frac{1}{10} + \frac{1}{15} = 0.3667$$

hence

$$R_{AD} = \frac{1}{0.3667} = 2.727 \ \Omega$$

The 10-Ω and 20-Ω parallel-connected resistors between the points D and E can be replaced by R_{DE} as follows

$$R_{DE} = \frac{(10 \times 20)}{(10 + 20)} = 6.667 \ \Omega$$

The resulting diagram is shown in Figure 3.14(b). In Figure 3.14(c), the series-connected resistors between A and C are combined to form R_{AP} having the value $(1 + 1.333) = 2.333 \ \Omega$, and the series-connected pair of resistors between A and E are replaced by R_{AE} having the value $(2.727 + 6.667) = 9.394 \ \Omega$.

In Figure 3.14(d), the parallel combination of resistors R_{AC} and R_{AE} are replaced by an equivalent resistor R_{ACE} having the value

$$R_{ACE} = \frac{(2.333 \times 9.394)}{(2.333 + 9.394)} = 1.869 \ \Omega$$

Finally, the resistance R_{AF} between points A and F is calculated as follows

$$R_{AF} = R_{ACE} + 40 \ \Omega = 41.869 \ \Omega$$

That is, a single 41.869-Ω resistor can be used to replace the circuit in Figure 3.14(a).

3.15 KIRCHHOFF'S LAWS

To determine the current flow in an electrical circuit, we need certain rules or laws which allow us to write down the circuit equations. Two laws laid down by Gustav Robert Kirchhoff, a German physicist, form the basis of all methods of electrical circuit solution.

First law

> *The total current flowing towards any junction of node in a circuit is equal to the total current flowing away from it.*

Second law

> *In any closed circuit or mesh, the algebraic sum of the potential drops and e.m.fs is zero.*

The *first law* is illustrated in Figure 3.15(a). The total current **flowing towards junction J** is $(I_1 + I_4)$, and the total current **flowing away from it** is $(I_2 + I_3)$. Hence, at junction J

$$I_1 + I_4 = I_2 + I_3$$

fig 3.15 *(a) Kirchhoff's first law and (b) Kirchhoff's second law*

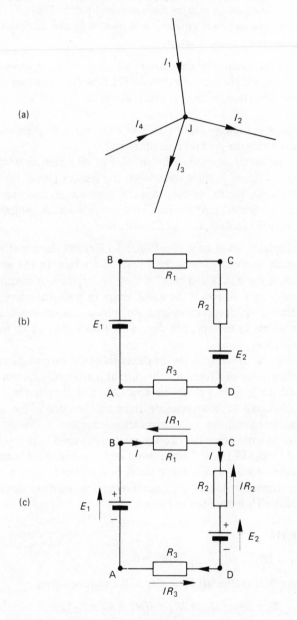

The *second law* is illustrated in Figure 3.15(b) and (c). In the **closed loop** ABCDA (remember, you *must* always return to the point from which you start in a *closed* loop), there are two e.m.fs (produced by the cells)

and three potential drops ('dropped' across the resistors). In order to write down the equation relating the e.m.fs and p.ds in Figure 3.15(b), it is necessary to add some information as shown in Figure 3.15(c); to do this the following steps must be taken:

1. Draw **on the circuit** the direction in which the current flows (if this is not known, simply *assume a direction of flow* [as will be seen later, the assumed direction is unimportant so far as the calculation is concerned]).
2. Draw an *e.m.f. arrow* by the side of each e.m.f. in the circuit which points towards the positive pole of the cell.
3. Draw a *potential drop arrow* by the side of each resistor which points towards the more positive terminal of the resistor (*Note*: the potential arrow *always* points in the opposite direction to the current flow through the resistor). At the same time, *write down the equation of the p.d. across the resistor*, that is, IR_1, IR_2, etc.

Having completed these steps (see Figure 3.15(c) and **starting at any point in the circuit**, proceed around the circuit and **return to the same point** writing down the e.m.fs and p.ds as they are reached. A *positive sign* is given to any e.m.f. or p.d. if the e.m.f. arrow or potential arrow points in the direction in which you move around the loop; a *negative sign* is given if the arrow points in the opposite direction. We will now apply these rules to Figure 3.15(b).

First, draw an arrow showing the direction of the current. Since you do not know the value of E_1 and E_2, the current is arbitrarily chosen to circulate around the loop in a clockwise direction. The direction in which the current is assumed to flow therefore 'fixes' the direction of the 'potential' arrows associated with the p.d. across each resistor.

In order to write down the circuit equation, we need to proceed around the closed loop ABCDA and write down each e.m.f. and p.d. according to the procedure outlined. We will traverse around the loop in a clockwise direction; starting at point A, the circuit equation is written down according to Kirchhoff's second law as follows.

Loop ABCDA

$$+E_1 - IR_1 - IR_2 - E_2 - IR_3 = 0$$

Collecting e.m.fs on the left-hand side of the equation gives

$$E_1 - E_2 = IR_1 + IR_2 + IR_3 = I(R_1 + R_2 + R_3)$$

You would find it an interesting exercise to write down the equation for the circuit starting at, say, point C and circulating around the loop in the opposite direction. The resulting equation should be the same as the one obtained above. (Since the circuit has not changed, the equation of the circuit should not change either!)

3.16 AN APPLICATION OF KIRCHHOFF'S LAWS

One form of procedure for applying Kirchhoff's laws, known as the **branch current method** is outlined below.

1. Assign a current to each branch of the circuit.
2. Apply Kirchhoff's first law to each junction in the circuit (take care to eliminate redundant currents! [see example below]).
3. Apply Kirchhoff's second law to each closed loop to give an equation for each loop (you need as many equations as you have unknown currents in step 2).
4. Solve the resulting equations for the unknown currents.

To illustrate this method, we will solve the electrical circuit in Figure 3.16. The circuit contains two batteries, battery 1 having an e.m.f. of 9 V and internal resistance 0.5 Ω, battery 2 having an e.m.f. of 6 V and internal resistance 0.25 Ω. The batteries are connected in parallel with one another, the two being connected to a 20-Ω load.

An inspection of the circuit shows that current flows in both batteries and in the load resistor. However, you will note that the current flowing in the load (the current from junction C to junction F) is the sum of the current from the two batteries (it is assumed for the moment that both batteries discharge into the load resistor). That is to say, there are only

fig 3.16 *solving an electrical circuit*

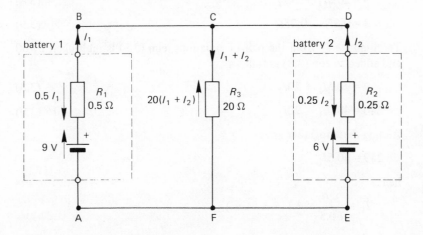

two unknown currents, namely I_1 and I_2 (the current in the 20-Ω resistor being equal to $I_1 + I_2$).

The circuit contains *three closed loops*, namely ABCFA, ABCDEFA and CDEFC. Any pair of loops can be used to give the equations we need, the first two being chosen in this case.

Loop ABCFA
The current directions selected are shown in Figure 4.16, and the e.m.f. and p.d. arrows are drawn according to the rules laid down earlier. Starting at point A and proceeding around the loop in the direction ABCFA, the loop equation is

$$9 - 0.5I_1 - 20(I_1 + I_2) = 0$$

or

$$9 = 20I_1 + 0.5I_1 + 20I_2$$
$$= 20.5I_1 + 20I_2 \tag{3.8}$$

Loop ABCDEFA
Starting at point A and proceeding around the loop in a clockwise direction gives the equation

$$9 - 0.5I_1 + 0.25I_2 - 6 = 0$$

or

$$3 = 0.5I_1 - 0.25I_2 \tag{3.9}$$

Equations (3.8) and (3.9) are collected together as follows

$$9 = 20.5I_1 + 20I_2 \tag{3.8}$$

$$3 = 0.5I_1 - 0.25I_2 \tag{3.9}$$

To eliminate I_2 from the pair of equations, eqn (3.9) is multiplied by 80 and added to eqn (3.8) as follows

$$9 = 20.5I_1 + 20I_2 \tag{3.8}$$

$$240 = 40.0I_1 - 20I_2 \qquad 80 \times (3.9)$$

Adding these together gives

$$249 = 60.5I_1$$

hence

$$I_1 = \frac{249}{60.5} = 4.116 \text{ A (Ans.)}$$

The value of I_2 is calculated by substituting the value of I_1 either into eqn (3.8) or eqn (3.9). We will use eqn (3.9) as follows:

$$3 = 0.5I_1 - 0.25I_2 = (0.5 \times 4.116) - 0.25I_2$$

$$= 2.058 - 0.25I_2$$

hence

$$0.25I_2 = 2.058 - 3 = -0.942$$

therefore

$$I_2 = \frac{-0.942}{0.25} = -3.768 \text{ (Ans.)}$$

The negative sign associated with I_2 merely implies that the current in battery 2 flows in the opposite direction to that shown in Figure 3.16. That is to say, battery 2 is 'charged' by battery 1!

The current in the 20-Ω load is calculated by applying Kirchhoff's first law to junction C (you could equally well use junction F for that matter). *You must be most careful always to use the value of I_2 calculated above*, that is, -3.768 A, *wherever I_2 appears in the circuit equations.*

$$\text{Current in the 20-ohm load} = I_1 + I_2$$

$$= 4.116 + (-3.768)$$

$$= 0.348 \text{ A (Ans.)}$$

The value of I_2 can be used to calculate the p.d. across the load as follows:

$$V_{CF} = 20(I_1 + I_2) = 20 \times 0.348 = 6.96 \text{ V (Ans.)}$$

SELF-TEST QUESTIONS

1. Describe the various types of fixed and variable resistors used in electrical engineering.
2. Why are 'preferred' values of fixed resistors used in low-power electrical circuits?
3. What is meant by the 'resistivity' of a material? Explain how the resistance of a conductor varies with its resistivity, length and cross-sectional area. Calculate the resistance in ohms of a 10-m length of copper at 20 °C whose diameter is 2 mm.
4. Calculate the conductance of the 10 m length of wire in question 3.3.
5. A heater coil has 500 turns of wire on it. The resistivity of the wire at its working temperature is 1.1 $\mu\Omega$m, and the coil is wound on a former of diameter 25 mm. Calculate the resistance of the wire at its working temperature if the cross-sectional area of the wire is 0.5 mm^2.

6. Three resistors of 10, 20 and 10 ohms, respectively, are connected (i) in series, (ii) in parallel. Determine the effective resistance of each combination. If the current drawn by each combination is 10 A, calculate (iii) the voltage across each combination and (iv) the power consumed in each case.

7. If, in Figure 3.16, V_1 = 15 V, V_2 = 20 V, R_1 = 10 Ω, R_2 = 20 Ω and R_3 = 30 Ω, calculate the current in each of the resistors.

SUMMARY OF IMPORTANT FACTS

A **resistor** may either be fixed or varaible. A range of **fixed resistors** are manufactured including **carbon composition, film resistors** and **wirewound resistors**. 'Light current' and electronics industries use a range of **preferred values** for fixed resistors whose values are coded in a **colour code**.

Variable resistors may either have a sliding contact or may be 'tapped' at various points along their length; they may be connected as **potentiometers** to provide a variable output voltage. A variable resistor can either have a **linear** resistance change with movement of the slider, or it can be **non-linear**, that is, logarithmic. The resistor track shape can have any one of several forms including **rectilinear, arc** or **multi-turn**.

The **resistance** of a resistor depends on several factors including the **resistivity**, the **length**, the **cross-sectional area** and the **temperature** of the material. The **conductance** of a conductor is equal to the *reciprocal of the resistance*.

In the case of a **conductor**, an increase in temperature causes an increase in resistance, and vice versa. In an **insulator** and a **semiconductor**, an increase in temperature causes a decrease in resistance.

When resistors are **connected in series**, *the resistance of the circuit is greater than the highest individual value of resistance in the circuit*. When they are connected in **parallel**, *the resistance of the circuit is less than the lowest individual value of resistance in the circuit*.

Electrical circuits can be solved by using **Kirchhoff's laws**. The **first law** states that the total current flowing towards a junction in the circuit is equal to the total current flowing away from the junction. The **second law** states that in any closed circuit the algebraic sum of the e.m.fs and p.ds is zero.

MEASURING INSTRUMENTS

AND ELECTRICAL

MEASUREMENTS

4.1 INTRODUCTION

So far we have discussed values of voltage, current, resistance, etc, without mentioning the way in which they are masured. In general, the operating principles of instruments are beyond the technical level reached at this point in the book and we are in the 'chicken and the egg' dilemma, insomuch that it is difficult to explain how an instrument works before its principle can be fully comprehended.

However, the way in which measuring instruments operate is so important that they are introduced at this point in the book. Several chapters devoted to electrical principles relevant to the operation of electrical measuring instruments are listed below. You should refer to these chapters for further information:

Electrostatics – Chapter 6
Electromagnetism – Chapter 7
Transformers – Chapter 14
Rectifiers – Chapter 16

4.2 TYPES OF INSTRUMENT

Instruments are classified as either analogue instruments or digital instruments. An **analogue instrument** – see Figure 4.1(a) – is one in which the magnitude of the measured electrical quantity is indicated by the movement of a pointer across the face of a scale. The indication on a **digital instrument** – see Figure 4.1(b) – is in the form of a series of numbers displayed on a screen; the smallest change in the indicated quantity corresponding to a change of ±1 digit in the *least significant digit* (l.s.d.) of the number. That is, if the meter indicates 10.23 V, then the actual voltage lies in the range 10.22 V to 10.24 V.

fig 4.1

(a) an analogue multimeter,   *(b) a digital multimeter*

Reproduced by kind permission of AVO Ltd

Both types of instrument have their advantages and disadvantages, and the choice of the 'best' instrument depends on the application you have in mind for it. As a rough guide to the features of the instruments, the following points are useful:

1. an analogue instrument does not (usually) need a battery or power supply;
2. a digital instrument needs a power supply (which may be a battery);
3. a digital instrument is generally more accurate than an analogue instrument (this can be a disadvantage in some cases because the displayed value continuously changes as the measured value changes by a very small amount);
4. both types are portable and can be carried round the home or factory.

4.3 EFFECTS UTILISED IN ANALOGUE INSTRUMENTS

An analogue instrument utilises one of the following effects:

1. electromagnetic effect;
2. heating effect;
3. electrostatic effect;
4. electromagnetic induction effect;
5. chemical effect.

The majority of analogue instruments including moving-coil, moving-iron and electrodynamic (dynamometer) instruments utilise the magnetic effect. The effect of the heat produced by a current in a conductor is used in thermocouple instruments. Electrostatic effects are used in electrostatic voltmeters. The electromagnetic induction effect is used, for example, in domestic energy meters. Chemical effects can be used in certain types of ampere–hour meters.

4.4 OPERATING REQUIREMENTS OF ANALOGUE INSTRUMENTS

Any instrument which depends on the movement of a pointer needs three forces to provide proper operation. These are:

1. a deflecting force;
2. a controlling force;
3. a damping force.

The **deflecting force** is the force which results in the movement or deflection of the pointer of the instrument. This could be, for example, the force acting on a current-carrying conductor which is situated in a magnetic field.

The **controlling force** opposes the deflecting force and ensures that the pointer gives the correct indication on the scale of the instrument. This could be, for example, a hairspring.

The **damping force** ensures that the movement of the pointer is *damped*; that is, the damping force causes the pointer to settle down, that is, be 'damped', to its final value without oscillation.

4.5 A GALVANOMETER OR MOVING-COIL INSTRUMENT

A **galvanometer** or **moving-coil instrument** depends for its operation on the fact that a current-carrying conductor experiences a force when it is in a magnetic field. A basic form of construction is shown in Figure 4.2. The 'moving' part of the meter is a coil wound on an aluminium *former* or frame which is free to rotate around a cylindrical soft-iron core. The moving coil is situated in the magnetic field produced by a permanent magnet; the function of the solf-iron core is to ensure that the magnetic field is uniformly distributed. The soft-iron core is securely fixed between the poles of the permanent magnet by means of a bar of non-magnetic material.

The moving coil can be supported either on a spindle which is pivoted in bearings (often jewel bearings) or on a *taut metal band* (this is the so-called **pivotless suspension**). The current, I, enters the 'moving' coil from the +ve terminal either via a spiral **hairspring** (see Figure 4.2) or via the

fig 4.2 *a simple galvanometer or moving-coil meter*

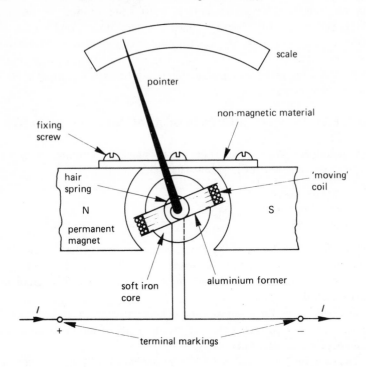

taut band mentioned above. It is this hairspring (or taut band) which provides the *controlling force* of the instrument. The current leaves the moving coil either by another hairspring or by the taut band at the opposite end of the instrument.

When current flows in the coil, the reaction between each current-carrying conductor and the magnetic field produces a mechanical force on the conductor; this is the *deflecting force* of the meter.

This force causes the pointer to be deflected, and as it does so the movement is opposed by the hairspring which is used to carry current into the meter. The more the pointer deflects, the greater the controlling force produced by the hairspring.

Unless the moving system is damped, the pointer will overshoot the correct position; after this it swings back towards the correct position. Without damping, the oscillations about the correct position continue for some time (see Figure 4.3 for the *underdamped response*). However, if the movement is correctly damped (see Figure 4.3) the pointer has an initial overshoot of a few per cent and then very quickly settles to its correct indication. It is the aim of instrument designers to achieve this response.

fig 4.3 *meter damping*

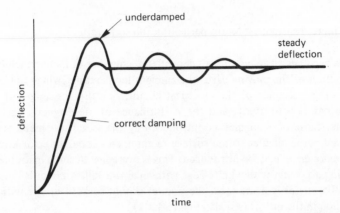

Damping is obtained by extracting energy from the moving system as follows. In the moving-coil meter, the coil is wound on an aluminium former, and when the former moves in the magnetic field of the permanent magnet, a current (known as an **eddy current**) is induced in the aluminium former. This current causes power to be consumed in the resistance of the coil former, and the energy associated with this *damps* the movement of the meter.

4.6 METER SENSITIVITY AND ERRORS

The **sensitivity** of an *ammeter* is given by the value of the current needed to give *full-scale deflection* (FSD). This may be 50 μA or less in a high quality analogue instrument.

The sensitivity of a *voltmeter* is expressed in terms of the *ohms per volt* of the instrument. That is

$$\text{voltmeter sensitivity} = \frac{\text{voltmeter resistance}}{\text{full-scale voltage}}$$

It can be shown that this is the reciprocal of the current needed to give full-scale deflection. A voltmeter having a sensitivity of 20 kΩ/V requires a current of

$$\frac{1}{20\,000} = 50 \times 10^{-6} \text{ A or } 50 \text{ } \mu\text{A}$$

to give full-scale deflection. The resistance of the voltmeter is therefore calculated from the equation

meter resistance = 'ohms per volt' × full-scale voltage

The **accuracy** of an instrument depends on a number of factors including the friction of the moving parts, the ambient temperature (which not only affects the resistance of the electrical circuit, but also the length of the hairsprings [where used], and the performance of the magnetic circuit). Another cause of error may be introduced by the user; if the meter is not observed perpendicular to the surface of the meter scale, an error known as *parallax error* may be introduced. This is overcome in some instruments by using a pointer with a knife-edge with an anti-parallax mirror behind it. When the meter is being read, the pointer should be kept in line with its reflection in the mirror (see also Figure 4.1(a)).

Strong magnetic fields can introduce errors in meter readings. These are largely overcome by using instruments which are housed in a magnetic shield which effectively deflects the external magnetic field from the meter.

Further errors can be introduced by using a meter of the 'wrong' resistance. For example, when an ammeter is inserted into a circuit, it increases the total resistance of the circuit by an amount equal to the resistance of the meter; in turn, this reduces the circuit current below its value when the ammeter is not in circuit, giving an error in the reading of current. Hence, *when an ammeter is used in a circuit, the resistance of the ammeter should be very much less (say $\frac{1}{100}$) of the resistance of the remainder of the resistance of the circuit into which it is connected.*

When a voltmeter is used to measure the p.d. across a component, *the resistance of the voltmeter must be much greater (say by a factor of 100) than the resistance of the component to which it is connected.* If this is not the case, the reading of the voltmeter differs from the voltage across the component when the voltmeter is not connected. For example, if you wish to measure the voltage across a 10 kΩ resistor, then the resistance of the voltmeter should be at least 100 × 10 kΩ = 1000 kΩ or 1 MΩ.

4.7 EXTENSION OF THE RANGE OF MOVING-COIL INSTRUMENTS

Moving-coil meters are basically low-current, low-voltage meters, and their circuits must be modified to allow them to measure either high current or high voltage. For example, the current needed to give full-scale deflection may be as little as 50 μA, and the voltage across the meter at FSD may be as little as 0.1 V.

Extending the current range

To extend the current range of the meter, a **low resistance shunt resistor**, S, is connected in parallel or in shunt with the meter as shown in Figure 4.4. The terms used in the diagram are

I = current in the external circuit to give FSD
I_G = current in the moving-coil meter to give FSD
R_G = resistance of the movement of the meter
I_S = resistance of the shunt resistor

The important equations for Figure 4.4 are

total current = meter current + shunt current

or

$$I = I_G + I_S \qquad (4.1)$$

and

p.d. across the meter = p.d. across the shunt

that is

$$I_G R_G = I_S S \qquad (4.2)$$

fig 4.4 *extending the current range of a meter using a shunt*

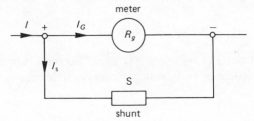

Example

The resistance of a moving-coil meter is 5 Ω and gives FSD for a current of 10 mA. Calculate the resistance of the shunt resistor required to cause it to have a FSD of 20 A.

Solution

Given that I = 20 A; I_G = 10 mA; R_G = 5 Ω from eqn (4.1

From eqn (4.1)

$$I_S = I - I_G = 20A - 10 \text{ mA}$$

$$= 20 - 0.01 = 19.99 \text{ A}$$

and from eqn (4.2)

$$\text{shunt resistance}, S = \frac{I_G R_G}{I_S} = (10 \times 10^{-3}) \times \frac{5}{19.99}$$

$$= 0.002501 \ \Omega$$

Extension of voltage range

The voltage range which can be measured by a moving-coil instrument can be increased by the use of a *voltage multiplier resistor*, R, which is connected in series with the meter as shown in Figure 4.5. Since the meter current, I_G, flows through both R and the meter, the voltage V_T, across the combination is

$$V_T = I_G(R + R_G) \tag{4.3}$$

fig 4.5 *extending the voltage range of a meter using a voltage multiplier resistor*

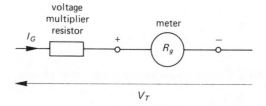

Example

A moving-coil meter of resistance 5 Ω gives FSD for a current of 10 mA. Calculate the resistance of the voltage multiplier which causes the meter to give a FSD of 100 V.

Solution

$$V_T = 100 \ \text{V}; I_G = 10 \ \text{mA}; R = 5 \ \Omega$$

From eqn (4.3)

$$V_T = I_G(R + R_G)$$

or

$$R = \frac{V_T}{I_G} - R_G$$

$$= \frac{100}{5 \times 10^{-3}} - 5 = 19\,995 \ \Omega$$

4.8 MEASUREMENT OF a.c. QUANTITIES USING A MOVING-COIL METER

A moving-ciol meter or galvanometer is only capable of measuring uni-directional (d.c.) current and voltage. However, it can be used to measure alternating current and voltage when the a.c. quantity has been converted into d.c. by means of a **rectifier** (see Chapter 16 for full details of rectifier circuits).

There are two basic types of rectifier circuit used with moving-coil meters, which are shown in Figure 4.6. In the 'half-wave' circuit in Figure 4.6(a), rectifier diode D1 allows current to flow through the meter when the upper terminal of the meter is positive with respect to the lower terminal; however, this diode prevents current flowing through the meter when the upper terminal is negative. It is the function of diode D2 to bypass the meter when the upper terminal is negative. In this way, the half-wave circuit allows current to flow through the meter **in one direction only** for one half of the a.c. cycle. No current flows through it in the other half-cycle.

The 'full-wave' rectifier circuit in Figure 4.6(b) allows current to flow through the meter from the terminal marked '+' to the terminal marked '−' whatever the polarity of the a.c. input.

Half-wave rectifiers are generally used in low-cost meters, and full-wave rectifiers are used with more expensive meters.

Extension of the a.c. range of moving-coil meters
The alternating current and voltage range of moving coil meters can, within limits, be extended using shunts and multiplying resistors, respec-tively, in the manner described earlier.

fig 4.6 *measurement of alternating current using (a) a half-wave circuit and (b) a full-wave circuit*

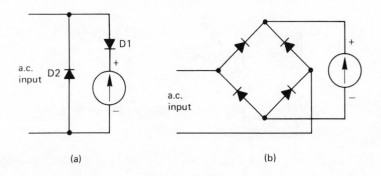

(a)   (b)

Very high values of current can be measured by using the meter in conjunction with a **current transformer** to reduce the measured current to, say, 5 A or less. Very high values of voltage can be measured by using the meter in conjunction with a **potential transformer** or **voltage transformer** to reduce the measured voltage level to, say, 110 V or less. The operating principles of the transformer are discussed in Chapter 14.

4.9 MOVING-IRON METERS

There are two basic types of meter in this category, namely

1. attraction-type meters;
2. repulsion-type meters.

Attraction-type meters depend on the fact that iron is attracted towards a magnetic field. Repulsion-type meters depends on the force of repulsion between two similarly magnetised surfaces (that is to say, there is a force of repulsion between two north poles, and there is a force of repulsion between two south poles).

The general construction of an **attractiont-ype moving-iron meter** is shown in Figure 4.7. The current, I, in the circuit flows through the coil and produces a magnetic field which attracts the 'moving iron' vane into the coil; this is the *deflecting force* of the meter. As the pointer moves it increases the tension in the controlling spring (this provides the *controlling force*). At the same time, it moves a piston inside a cylinder; the piston provides the *damping force* of the meter.

fig 4.7 *attraction-type moving-iron meter*

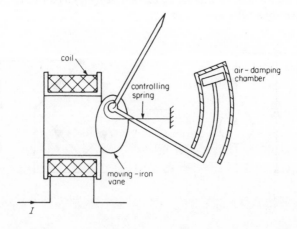

The basis of a **repulsion-type moving-iron meter** is shown in Figure 4.8. The *deflecting force* is again provided by the current in the coil, the *controlling force* by the controlling spring, and the *damping force* by the air dashpot. The method of operation is as follows. The instrument has two parallel iron rods or vanes inside the coil, one being fixed to the coil whilst the other is secured to the spindle and is free to move. When current flows in the coil, both rods are similarly magnetised, each having N-poles at (say) the upper end in the diagram and S-poles at the lower end. The like magnetic poles mutually repel one another and, since one is fixed to the coil, the rod which is attached to the spindle moves away from the fixed rod, causing the pointer to deflect.

Moving-iron instruments are generally more robust and cheaper than moving-ciol meters, but have a lower accuracy. They are widely used as general-purpose panel meters.

fig 4.8   *repulsion-type moving-iron meter*

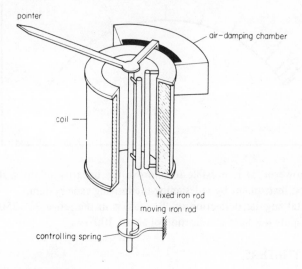

4.10 METER SCALES

The length and way in which an instrument scale is calibrated depends on a number of factors and, in general, results in one of two types of scale, namely a linear scale or a non-linear scale. A **linear scale** is one which has an equal angular difference between points on the scale (see Figure 4.9(a)), and a **non-linear scale** has unequal angular differences (Figure 4.9(b)).

A moving-coil meter has a linear scale for both current and voltage scales. A moving-iron meter has a non-linear scale for both voltage and

fig 4.9 *instrument scales: (a) linear, (b) non-linear*

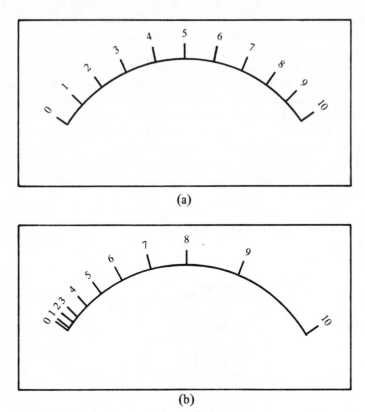

(a)

(b)

current; however, it is possible to improve the linearity of the scale of a moving-iron instrument by redesigning the moving vane system.

The total angular deflection of the scale is in the range 90°–250° (the scales in Figure 4.9 have a deflection of about 100°).

4.11 WATTMETERS

As the name of this instrument implies, its primary function is to measure the power consumed in an electrical circuit. The wattmeter described here is called an **electrodynamic wattmeter** or a **dynamometer wattmeter**. It is illustrated in Figure 4.10, and has a pair of coils which are fixed to the frame of the meter (the **fixed coils**) which carry the main current in the circuit (and are referred to as the **current coils**), and a **moving coil** which is pivoted so that it can rotate within the fixed coils. The moving coil generally has a high resistance to which the supply voltage is connected and is called the **voltage coil** or **potential coil**. The pointer is secured to the

fig 4.10 *Dynamometer wattmeter*

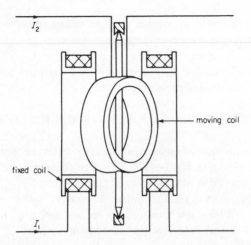

spindle of the moving coil. The magnetic flux produced by the fixed coils
is proportional to current I_1 in the figure, and the magnetic flux produced
by the moving coil is proportional to I_2 (the former being proportional to
the load current and the latter to the supply voltage; see Figure 4.11).

The two magnetic fields react with one another to give a deflecting
force proportinal to the product $I_1 \times I_2$. However, since I_2 is proportional
to the supply voltage V_S, the deflection of the meter is given by

meter deflection = $V_S I_1$ = power consumed by the load

fig 4.11 *wattmeter connections*

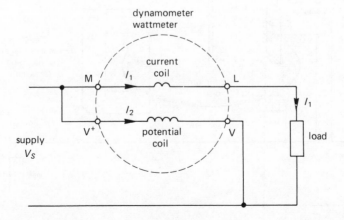

Dynamometer wattmeters can measure the power consumed in either a d.c. or an a.c. circuit.

Hairsprings are used to provide the *controlling force* in these meters, and air-vane damping is used to *damp* the movement.

The power consumed by a three-phase circuit (see Chapter 13) is given by the sum of the reading of two wattmeters using what is known as the **two-wattmeter method** of measuring power.

4.12 THE ENERGY METER OR KILOWATT–HOUR METER

The basic construction of an electrical energy meter, known as an **induction meter**, is shown in Figure 4.12. This type of meter is used to measure the energy consumed in houses, schools, factories, etc.

The magnetic field in this instrument is produced by two separate coils. The 'current' coil has a few turns of large section wire and carries the main current in the circuit. The 'voltage' coil has many turns of small section wire, and has the supply voltage connected to it. The 'deflection' system is simply an aluminium disc which is free to rotate continuously (as you will see it do if you watch your domestic energy meter), the disc rotating faster when more electrical energy is consumed.

fig **4.12**   *induction-type energy meter*

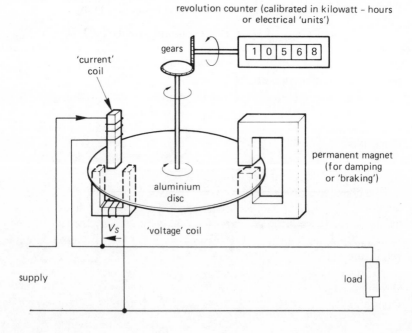

revolution counter (calibrated in kilowatt – hours or electrical 'units')

gears

| 1 | 0 | 5 | 6 | 8 |

'current' coil

permanent magnet (for damping or 'braking')

aluminium disc

V_S

'voltage' coil

supply

load

The effect of the magnetic field produced by the coils is to produce a torque on the aluminium disc, causing it to rotate. The more current the electrical circuit carries, the greater the magnetic flux produced by the 'current' coil and the greater the speed of the disc; the disc stops rotating when the current drawn by the circuit is zero.

The disc spindle is connected through a set of gears to a 'mileometer' type display in the case of a digital read-out meter, or to a set of pointers in some older meters. The display shows the total energy consumed by the circuit.

The rotation of the disc is *damped* by means of a permanent magnet as follows. When the disc rotates between the poles of the permanent magnet, a current is induced in the rotating disc to produce a 'drag' on the disc which damps out rapid variations in disc speed when the load current suddenly changes.

These meters are known as **integrating meters** since they 'add up' or 'integrate' the energy consumed on a continual basis.

4.13 **MEASUREMENT OF RESISTANCE BY THE WHEATSTONE BRIDGE**

The resistance of a resistor can be determined with high precision using a circuit known as a **Wheatstone bridge** (invented by Sir Charles Wheatstone).

A basic version of the Wheatstone bridge is the familiar 'slide-wire' version shown in Figure 4.13(a). The circuit consists of two parallel branches PQR and PSR supplied by a common source of e.m.f., E. Branch PSR is a wire of uniform cross-sectional area and of length typically 50 or 100 cm; for reasons given later, the two arms l_1 and l_2 are known as the **ratio arms** of the bridge. Branch PQR contains two series-connected resistors R_U and R_V, resistor R_U is the **unknown value of resistance** (to be determined by the measurements), and R_V is a **standard resistor** of precise value and having a high accuracy.

Point Q in the upper branch is connected to the **sliding contact**, S, on the slide-wire via a **centre-zero galvanometer**.

The circuit diagram of the bridge is shown in Figure 4.13(b), in which resistor R_1 replaces the length of slide-wire l_1, and R_2 replaces l_2. The circuit basically comprises two parallel-connected branches (one containing R_U and R_V and the other containing R_1 and R_2) which are linked by the galvanometer G.

When the circuit is initially connected to the battery, the current I_U flowing in R_U does not necessarily equal I_V which flows in R_V; the difference in value between these two currents flows in the galvanometer G.

fig 4.13 *(a) 'slide-wire' version of the Wheatstone bridge, (b) circuit diagram*

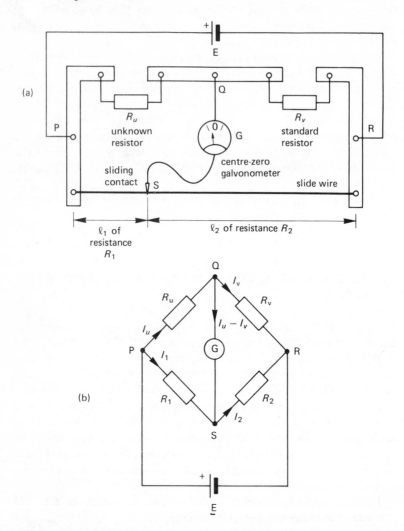

Depending on which way the current flows, the needle of the galvanometer will deflect either to the left or to the right of its centre-zero position.

The Wheatstone bridge at balance

If the slider is moved, say, to the left of its initial position the deflection of the valvanometer will (say) give a larger deflection. If the slider is moved to the right, the deflection decreases. By moving the slider along the slide-wire, a position will be reached when the galvanometer gives zero deflection, indicating that the current through it is zero. When this happens, the bridge is said to be **balanced**. Since the current in the galvanometer is zero, then

$$I_U = I_V \qquad (4.4)$$

and

$$I_1 = I_2 \qquad (4.5)$$

Now, since no current flows through the galvanometer, there is no p.d. across it, hence

potential at Q = potential at S

Furthermore it follows that

p.d. across R_U = p.d. across R_V

or

$$I_U R_U = I_1 R_1 \qquad (4.6)$$

and

p.d. across R_V = p.d. across R_2

or

$$I_V R_V = I_2 R_2 \qquad (4.7)$$

Dividing eqn (4.6) by eqn (4.7) gives

$$\frac{I_U R_U}{I_V R_V} = \frac{I_1 R_1}{I_2 R_2} \qquad (4.8)$$

But, at balance $I_V = I_U$ and $I_1 = I_2$, so that eqn (4.8) becomes

$$\frac{I_U R_U}{I_U R_V} = \frac{I_1 R_1}{I_1 R_2}$$

or

$$\frac{R_U}{R_V} = \frac{R_1}{R_2} \qquad (4.9)$$

The **unknown value of resistance** R_U can therefore be calculated as follows

$$R_U = R_V \times \frac{R_1}{R_2} \qquad (4.10)$$

In the above equation, R_1 and R_2 correspond to the resistance of l_1 and l_2, respectively, of the slide-wire. Since R_1 and R_2 in eqn (4.10) form a resistance ratio, these two 'arms' of the bridge are known as the **ratio arms**.

Unfortunately we only know the length of l_1 and l_2 and do not know their resistance. However, since the slide-wire has a uniform cross-sectional area and each part has the same resistivity, eqn (4.10) can be rewritten in the form

$$R_U = R_V \times \frac{l_1}{l_2} \tag{4.11}$$

Example
The value of an unknown resistance R_U is measured by means of a Wheatstone bridge circuit of the type in Figure 4.13(a). Balance is obtained when $l_1 = 42$ cm, the total length of the slide-wire being 100 cm. If the value of the standard resistor is 10 Ω, determine the value of R_U.

If the unknown value of resistor is replaced by a 15 Ω resistor, and a standard resistor of 20 Ω is used, what is the value of l_1 which gives balance?

Solution
For the first part of the question $R_V = 10\ \Omega, l_1 = 42$ cm, $l_2 = (100 - 42) = 58$ cm
From eqn (4.11)

$$R_U = \frac{R_V l_1}{l_2} = 10 \times \frac{42}{58} = 7.24\ \Omega \text{ (Ans.)}$$

For the second part of the problem

$$R_U = 15\ \Omega, R_V = 20\ \Omega$$

From eqn (4.11)

$$\frac{l_1}{l_2} = \frac{R_U}{R_V} = \frac{15}{20} = 0.75$$

or

$$l_2 = \frac{l_1}{0.75} = 1.333\, l_1.$$

But $l_1 + l_2 = 100$ cm hence $l_1 + 1.333\, l_1 = 100$ cm
therefore

$$l_1 = \frac{100}{(1 + 1.333)} = 42.86 \text{ cm (Ans.)}$$

4.14 MEASUREMENT OF RESISTANCE USING AN OHMMETER

An ohmmeter is a moving-coil instrument calibrated directly in ohms. Basically it is a galvanometer fitted with a battery E (see Figure 4.14(a)) to which the unknown resistor R is connected. The current which flows through the meter when the unknown resistance is connected is a measure of the resistance of the resistor; the lower the value of R the larger the current, and vice versa.

When the unknown resistance is disconnected, the meter current is zero; the ohms scale is therefore scaled to show **infinite (∞) ohms when the current is zero**. The meter is calibrated by applying a short-circuit to its terminals, and then adjusting the SET ZERO control resistor RV until the needle gives full-scale deflection. That is, **maximum current corresponds to zero ohms**.

Since the current in the unknown resistance is proportional to $\frac{1}{R}$, the scale calibration is non-linear (see Figure 4.14(b)).

fig 4.14 *(a) typical ohmmeter circuit and (b) an example of its scale*

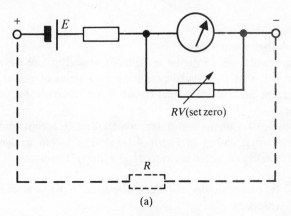

(a)

(b)

4.15 ELECTRONIC INSTRUMENTS

These are divided into two gategories, namely analogue instruments and digital instruments.

Electronic analogue instruments

These consist of a moving-coil meter which is 'driven' by an amplifier as shown in Figure 4.15. The function of the amplifier is not only to amplify a voltage or a current signal which may have a very low value, but also to give the instrument a high 'input resistance'. The latter means that when the instrument is connected to the circuit, it draws only a minute current from the circuit.

fig 4.15 *the basis of an analogue electronic instrument*

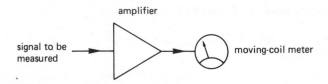

Electronic digital instruments

There are two types of digital instrument, namely those which accept a digital signal, that is, a signal consisting of a series of pulses, and those which accept an analogue signal (similar to a conventional voltmeter or ammeter).

The basis of a digital instrument which directly accepts pulses from a digital system is shown in Figure 4.16(a). The incoming signal may be from an installation which is electronically 'noisy' (an electrical motor or a

fig 4.16 *(a) timer/counter, (b) block diagram of a digital ammeter or voltmeter*

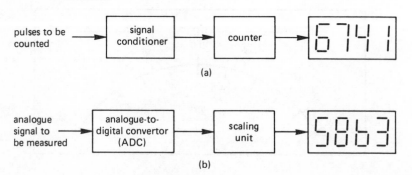

fluorescent lamp are electrically 'noisy' devices); it is first necessary to 'condition' the incoming signal not only to remove the 'noise' but also to bring it to the correct voltage level for the meter. The pulses are then counted and displayed on a suitable read-out device; typical display devices include **light-emitting diodes** (LED), **liquid crystal displays** (LCD), **gas-discharge tubes** and **cathode-ray tubes** (CRT).

The basis of an instrument which measures an analogue quantity such as voltage or current, but provides a digital display of its value is shown in Figure 4.16(b). The analogue quantity is first converted to its digital form by means of an **analogue-to-digital convertor** (ADC). The output from this part of the circuit is in digital (or on–off) form, the digital signal being related to the analogue value being measured. The digital signal is then displayed on a suitable device. The measuring range of such a meter can be extended using either voltage multiplier resistors or current shunts, in the manner outlined for conventional analogue meters.

4.16 MEASUREMENT OF RESISTANCE USING A DIGITAL METER

The basis of a digital ohmmeter is shown in Figure 4.17. It consists of a digital voltmeter of the type in Figure 4.16(b), together with a constant current source, both being housed in the ohmmeter.

When an unknown resistor R is connected to the terminals of the instrument, the constant value of current (I) from the 'current' source flows through R to give a p.d. of IR at the terminals of the voltmeter. This voltage is measured by the electronic voltmeter and, since the p.d. is pro-

fig 4.17 *principle of a digital ohmmeter*

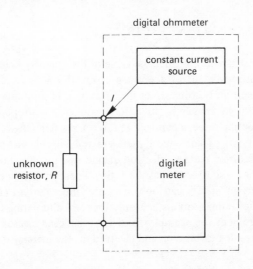

portional to the resistance of the unknown resistor, the measured voltage is proportional to the resistance of R. The voltage displayed on the meter can therefore be calibrated in 'ohms'.

SELF-TEST QUESTIONS

1. Describe the effects which can be utilised to provide a reliable electrical measuring instrument. Given an example of each kind of effect.
2. What are the basic requirements of an analogue measuring instrument? How are these needs provided in (i) a moving-coil meter, (ii) a moving-iron meter, (iii) an electrodynamic instrument and (iv) an induction instrument?
3. Explain the operation of a moving-coil meter and discuss the way in which the movement is damped.
4. Why can electrodynamic meters be used to measure the power consumed by an electrical circuit?
5. Draw a circuit diagram of a Wheatstone bridge and determine an equation which gives its 'condition of balance'.
6. The value of a resistor of 40-Ω resistance is measured using a Wheatstone bridge. If the value of the standard resistor is 20 Ω and the length of the slide-wire is 100 cm, determine the position of the slider.
7. Draw a circuit diagram of an ohmmeter and explain its operation. Why does the 'zero ohms' position of the scale correspond to the 'maximum current' position?
8. Discuss the operating principles of electronic instruments.
9. Explain how the range of (i) a moving-coil meter and (ii) a moving-iron meter can be extended to measure a large current and a large voltage.

SUMMARY OF IMPORTANT FACTS

An **analogue instrument** indicates the value of the quantity being measured by the smooth movement of a pointer across the scale of the meter; a **digital instrument** gives a display in the form of a set of numbers or digits.

An *analogue meter* needs three forces to ensure proper operation, namely, a **deflecting force**, a **controlling force** and a **damping force**.

Important types of analogue meters are **moving-coil meters (galvanometers)**, **moving-iron meters**, **electrodynamic instruments** and **induction instruments**. The scales used may be either **linear** or **non-linear**. In their basis form, certain meters can only measure direct current (the moving-coil meter for example), others can only measure alternating current (the induction instrument for example), and yet others can measure either d.c. or a.c. (for example moving-iron and electrodynamic meters).

The **range** of a meter can be extended to measure a higher current using a current **shunt**, or to measure a higher voltage using a series-connected **voltage multiplier** resistor.

An *unknown value of resistance* can be measured using a **Wheatstone bridge**; the bridge uses an accurate **standard resistor** together with the two **ratio arms** of the bridge. Resistance can also be measured using an **ohmmeter**.

Electronic instruments are subdivided into **analogue indication** and **digital indication** types. *Analogue indication* types generally consist of a conventional moving-coil meter which is 'driven' by an amplifier. *Digital indication* types are generally more complex than analogue indication types.

ELECTRICAL ENERGY AND ELECTRICAL TARIFFS

5.1 **HEATING EFFECT OF CURRENT: FUSES**

The amount of heat energy, W joules or watt seconds, produced by a current of I amperes flowing in a resistor R ohms for t seconds is

$W = I^2Rt$ joules (J) or watt seconds

Thus, if 1 A flows for 1 second in a resistance of 1 Ω, the energy dissipated is

$W = I^2 \times 1 \times 1 = 1$ J

However, if 100 A flows for the same length of time in the same resistor, the energy dissipated is

$W = 100^2 \times 1 \times 1 = 10\,000$ J

The electrical power rating of an item of electrical plant is related to its ability to dissipate the energy which is created within the apparatus. Since the heat energy is related to I^2, the **rating of an electrical machine** depends on its ability to dissipate the heat generated (I^2R) within it; the rating therefore depends **on the current which the apparatus consumes.**

In order to protect electrical equipment from excessive current, an electrical **fuse** is connected in series with the cable supplying the plant. The function of the fuse is to 'blow' or melt when the load-current exceeds a pre-determined value; needless to say, the current-carrying capacity must be less than that of the cable to which it is connected!

The fuse is simply a piece of wire (which may be in a suitable 'cartridge') which, when carrying a 'normal' value of current remains fairly cool. However, if the current in the circuit rises above the 'rating' of the fuse, the heat generated in the fuse rises to the point where the melting-point of the fuse material is reached. When the fuse melts, the circuit is broken, cutting off the current to the faulty apparatus.

When wiring a fuse, you should always ensure that it is connected in the 'live' wire; this ensures that, when the fuse 'blows', the apparatus is disconnected from the live potential.

In general, the larger the current passing through the fuse, the shorter the 'fusing' time; that is, a fuse has an **inverse-time fusing characteristic**. That is to say, a current which is only slightly greater than the current rating of the fuse may take many minutes to blow the fuse, but a current which is, say, ten times greater than the fuse-rating causes it to blow in a fraction of a second.

Early types of fuses could be rewired, that is, the fuse was simply a piece of wire connected between two terminals in a ceramic fuseholder. The rating of the fuse-wire in a domestic system would be, typically, 5 A, 15 A or 30 A. Modern domestic apparatus is protected by means of a fuse in the plug which terminates the cable. This type of fuse is known as a **cartridge fuse**, which is a cartridge containing the fusible element; the rating of these fuses in the UK is usually one of the values 3 A, 5 A, 10 A or 13 A.

Industrial fuses handle very high values of current (typically many hundreds of amperes) and are frequently special fuses known as **high rupturing-capacity fuses** or **high breaking-capacity fuses**.

Many items of electronic apparatus have fuses in which the fusible element is housed in a glass cartridge, allowing the user to inspect the state of the fuse when it is withdrawn from the fuseholder. Certain types of electronic apparatus draw a sudden rush of current when switched on, but take a much lower current when the apparatus has 'settled down': in this case, a special **anti-surge fuse** is used. This type of fuse tolerates a large rush of current for a few seconds, after which it protects the circuit normally.

5.2 CALCULATION OF ELECTRICAL ENERGY

The basic unit of electrical energy is the joule, and *one joule of energy is consumed when one watt of power is absorbed for a time of one second.* That is

energy, W = power (watts) × time (seconds) J

The joule or watt–second is a very small unit of energy, and the unit used to measure energy in domestic and industrial situations is the **kilowatt-hour** (kWh). The type of meter used to measure energy using this unit is the *kilowatt–hour meter*, and is shown in Figure 4.12 of Chapter 4. The energy consumed in kWh is calculated from the equation

energy in kilowatt-hours = power (kW) × time (h) kWh

If a 3 kW electric fire is switched on for 4 hours, the energy consumed is

energy, $W = 3 \times 4 = 12$ kWh

By comparison, the energy consumed in joules is

energy, $W = 3000 \times (4 \times 60 \times 60) = 43\,200\,000$ J

The relationship between joules and kilowatt hours is

1 kWh $= 1$ (kW) $\times 1$ (h) $= 1000$ (W) $\times (60 \times 60)$ s

$= 3\,600\,000$ J $= 3.6$ MJ

The power, P, consumed by a resistor R which carries a current of I ampere is given by

power, $P = I^2R = \dfrac{V^2}{R} = VI$ W

where V is the voltage across the resistor. The energy, W, consumed in joules by the resistor in t seconds is

energy, $W = I^2Rt = \dfrac{V^2t}{R} = VIt$ J

Also, the power consumed in kW in the resistor is

power, $P = \dfrac{I^2R}{1000} = \dfrac{V^2}{1000R} = \dfrac{VI}{1000}$ kW

and the energy consumed in kWh in T hours is

$W = \dfrac{I^2RT}{1000} = \dfrac{V^2T}{1000R} = \dfrac{VIT}{1000}$ kWh

Example
Calculate the amount of energy consumed in (i) joules, (ii) kWh by a 10-Ω resistor which is connected to a 9-V battery for 20 seconds.

Solution:

$R = 10\ \Omega, E = 9$ V, $t = 20$ s

(i) Energy in joules $= E^2tR = 9^2 \times \dfrac{20}{10} = 162$ J (Ans.)

(ii) Energy in kWh $= \dfrac{\text{joules}}{3.6 \times 10^6}$

$= 45 \times 10^{-6}$ kWh or units of electricity.

Example

Calculate the amount of energy consumed (i) in joules, (ii) in kWh by a 100-W electric lamp which is switched on for 5 hours.

Solution

$P = 100$ W, $T = 5$ h = 5×3600 s = $18\,000$ s

(i) Energy in joules = power (W) × time (s)

= $100 \times 18\,000 = 1\,800\,000$ J

(ii) Energy in kWh = power (kW) × time (h)

= $\dfrac{100}{(1000)} \times 5 = 0.5$ kWh or units of electricity

5.3 APPLICATIONS OF HEATING EFFECTS

Many domestic and industrial applications depend on electro-heat for their operation. The simplest of the devices in this category is the **bimetallic strip** consisting of two metals which are in close contact with one another, but have different coefficients of expansion when heated (see Figure 5.1). When the strip is cold, both metals are the same length (see Figure 5.1(a)). When current passes through the strip both metals heat up but, because of their differing coefficients of expansion, the brass expands more rapidly than the iron (Figure 5.1(b)). Since one end of the strip is secured, the 'free' end bends upwards. This principle has a number of applications.

One application is in the form of a **thermal overcurrent trip** as shown in Figure 5.1(c). If the current drawn by, say, a motor passes through the bimetallic strip (or, more likely, through a heating coil wrapped round the strip), the effect of the load current is to heat up the strip. Under normal conditions of electrical load, the strip does not get hot enough to bend and break the contacts at the free end of the strip. However, when the motor is subject to a heavy overload, the heavy current heats the strip sufficiently to cause the strip to bend upwards and break the contacts; this cuts off the current to the overloaded motor, and prevents it from being damaged. A practical current overload trip should, of course, have a means of 'latching' the contacts open until they are 'reset' (closed) by a qualified engineer.

Alternatively, the bimetallic strip can be used as a temperature-measuring element in a central-heating control system in a house, that is it can be used as a **thermostat**. In this case the bimetallic strip is not heated by an electrical current, but by the room temperature (or boiler temperature). When the room (or boiler) is cool, the bimetallic strip is fairly straight and the contacts shown in inset (i) are closed. The current I_1 flowing through

fig 5.1 *bimetallic strip: (a) when cold, (b) when hot, (c) typical
applications*

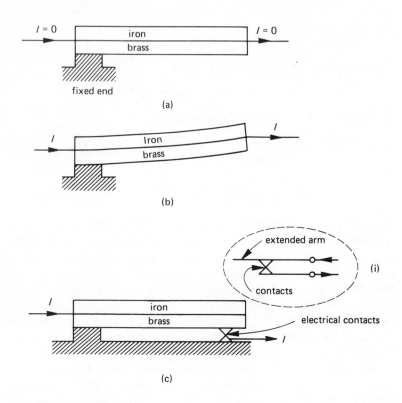

(a)

(b)

(c)

the contacts operates the central heating system and, so long as the contacts are closed, the boiler heats the system up. However, as the room heats
up the bimetallic strip begins to bend upwards. When the room reaches the
required temperature, the free end of the strip breaks the contacts and
cuts off current I_1; this has the effect of turning the central heating off.
When the room cools down, the bimetallic strip straightens out and the
contacts close again; at this point the central heating system turns on
again.

A thermostat has a manual control on it, the function of this control
being to alter the separation between the strip and the contacts (a larger
gap gives a higher room temperature, and a smaller gap giving a lower
temperature).

5.4 ELECTRICITY SUPPLY TARIFFS

The way in which any user of electricity is charged for electricity depends on a number of factors including:

1. the type of consumer they are, for example, whether the electricity is consumed in a house, a factory, a farm, or a place of worship, etc.;
2. the time of the day during which electricity is consumed, for example, all day, or part of the day, or at night, etc.

The simplest form of **electricity tariff** is one where the consumer is charged a fixed amount for each unit of electrical energy (kWh) consumed. Unfortunately, this type of tariff penalises the consumer who uses large amounts of electricity (and is therefore a good customer). The most popular type of tariff for domestic and small industrial consumers is the **two-part tariff**; the two parts to the tariff are

Part 1: this is related to the consumer's contribution to the **standing charge** (see below) of the production of electricity

Part 2: this is related to the consumer's contribution to the **running charges** (see below) of the production of electricity.

Standing charges
These charges are independent of the actual cost of producing electricity, and include the money needed to build the power station and supply lines, together with the rent, rates, telephone bills, etc, and the salaries of the administrative staff. It also includes an amount of money to cover the depreciation of the plant, so that it can be renewed at some later date.

Running charges
These charges are directly related to the electrical output of the power station and include items of money for such things as fuel, stores, repairs, energy loss in the transmission system, operatives wages, etc.

5.5 MAXIMUM DEMAND

The 'maximum demand' that a consumer places on the electricity supply is of interest to the generating authority, since this influences the size of the generating plant he has to install. In general, the maximum demand (MD) is mainly of interest to large industrial consumers, and is the highest value of volt-amperes consumed at the premesis. You should note that whilst the volt–ampere (VA) product in a d.c. circuit is equal to the electrical power consumed in watts, it is not necessarily equal to the power consumed in an a.c. circuit (for details see the chapters on a.c. circuits); for this reason we refer to it as the VA product rather than the power consumed.

Clearly, a consumer who has a high MD consumes a large amount of electricity. It is usually the case that the supply tariff arranges for a large consumer of electricity to pay progressively less for each kilo-volt-ampere (kVA or 1000 VA) or maximum demand he consumes. This is illustrated in section 5.6.

5.6 A TYPICAL SUPPLY TARIFF

The equation of the electricity supply tariff for a domestic or small industrial premises can be written in the form

Cost = £X per month (or quarter, or annum) as a standing charge
+ Y pence per unit of electricity

For example, a typical domestic tariff may be

a standing charge per quarter of £6.50
a cost per unit of electrical energy of 6.5p

For a large industrial consumer the tariff may take the form

Cost = £L per month as a standing charge

+ £M per kVA of maximum demand (MD) for the first, say, 100 kVA

+ £N per kVA of MD for the next 200 kVA, etc.

A typical industrial tariff may be

a standing charge per month of £20
an additional charge for each kVA of MD
for the first 100 kVA of £1.00
for the next 200 kVA of £0.75
for each additional kVA of £0.50

The industrial consumer may also have to pay an additional 'meter' charge (depending on the voltage at which the energy is metered). He may have to pay a 'fuel cost' charge, which depends on the cost of the fuel purchased by the supply authority.

5.7 ELECTRICITY BILLS

The foregoing indicates that the calculation of the cost of electrical energy can sometimes be complicated. In the following an example involving a domestic installation is calculated.

Example

A family consume 800 units of electrical energy per quarter. If the supply tariff consists of a standing charge of £6.50 and a cost per unit of 6.5p, calculate the cost of electrical energy during the quarter.

Solution

The cost of the electrical energy actually consumed is

$$\frac{£(\text{number of units} \times \text{cost in pence per unit})}{100}$$

$$= \frac{£800 \times 6.5}{100} = £52.00$$

The final bill is

Cost per quarter $= £(\text{standing charge} + \text{cost of electricity})$

$$= £(6.50 + 52.00) = £58.50$$

SELF-TEST QUESTIONS

1. Calculate the energy consumed in MJ and kWh when a current of 20 A flows in a resistor of 100 Ω for 1 hour.
2. Describe the meaning of (i) a fuse, (ii) a high rupturing-capacity fuse, (iii) an anti-surge fuse.
3. Explain what is meant by a two-part electricity tariff, and state the reason for each 'part' of the tariff.
4. Why do industrial consumers need to pay a 'maximum demand' charge?

SUMMARY OF IMPORTANT FACTS

Electrical equipment can be protected by a **fuse** which, in its simplest form, is a wire which melts when a current in excess of its rated current flows through it for a short length of time.

The unit of **electrical energy** (power × time) is the joule, but a more practical unit is the **kilowatt-hour** (kWh).

An electricity **tariff** is the way in which you are charged for electricity. Tariffs are many and varied, one of the most popular being the **two-part tariff**. One part of this tariff relates to the **standing charges** associated with production of electricity, the second part relates to the **running charges**. Many industrial consumers have to pay a **maximum demand** charge.

CHAPTER 6

ELECTROSTATICS

6.1 FRICTIONAL ELECTRICITY

Many hundreds of years BC it was discovered that when amber was rubbed with fur, the amber acquired the property of being able to attract other objects. The reason for this was not understood until man knew more about the structure of matter. What happens when, for example, amber and fur are rubbed together is that electrons transfer from one to the other, with the result that the charge neutrality of the two substances is upset. The substance which gains electrons acquires a negative charge, and the one which loses electrons acquires a positive charge.

It was also found that the effects of the charge produced by friction can only be observed on a good insulating material, and that the charge itself can only be sustained in dry conditions. These facts are in accord with our knowledge of electricity since, under damp conditions, electrical charge 'leaks' away from charged bodies. This knowledge is used today by aircraft manufacturers; when an aircraft is in flight, it can build up a high frictional charge of electricity, but using tyres containing conducting substances the charge is allowed to leak away when the aircraft lands.

If you rub a polythene rod with a woollen duster, the polythene acquires a negative charge; when a glass rod is rubbed with silk, the rod acquires a positive charge. When you have a charged rod, you can make some interesting observations as illustrated in Figure 6.1.

When two negatively charged polythene rods are brought into close proximity with one another, as in Figure 6.1(a), and if one of the rods is suspended on a thread, there is a force of repulsion between the fixed and suspended rods. However, if a positively charged glass rod is brought close to a negatively charged polythene rode – see Figure 6.1(b) – there is a force of attraction between the rods. This is summarised as follows:

Like electrical charges repel one another.
Unlike electrical charges attract one another.

fig 6.1 *electrostatics: (a) like charges repel, (b) unlike charges attract*

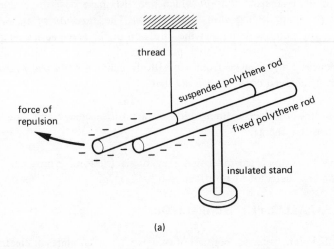

(a)

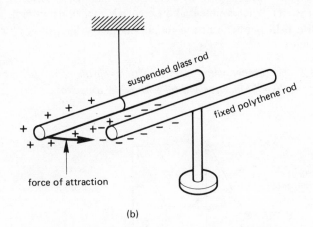

(b)

6.2 **THE UNIT OF ELECTRICAL CHARGE**

The charge stored on an electrical conductor (whether produced by friction or other means) is measured in **coulombs** (unit symbol C).

6.3 **ELECTRIC FLUX**

A charged electrical body is said to be surrounded by an **electric field**, and a measure of the field is its **electric flux** (symbol Q) having units of the coulomb. *One unit of electrical charge* produces *one unit of electric flux*; that is, 1 coulomb of electric charge produces 1 coulomb of flux. Quite often we think of an electric charge travelling along a *line of flux*; this 'line' is an imaginary concept which scientists have devised to simplify problems when dealing with electrostatics. The *direction of an electric field* at any point is defined as the direction of the force acting on a *unit positive charge* placed at that point. Clearly, since a positive charge experiences a force *away* from a positively charged body and *towards* a negatively charged body, we can say that

a 'line' of electric force starts on a positive charge and ends on a negative charge.

However, you must remember that a line of electric force is imaginary, and you can only detect it by its effect on other bodies.

6.4 A PARALLEL-PLATE CAPACITOR

A parallel-plate capacitor consists of two parallel metal **plates** or **electrodes** (see Figure 6.2) separated by an insulating material or **dielectric** (in Figure 6.2 it is air). When a potential difference is applied between the plates, an **electric field** is established in the dielectric and lines of electric flux link the plates.

fig 6.2   *(a) parallel-plate capacitor with an air dielectric; (b) symbol for a 'fixed' capacitor and (c) a 'variable' capacitor*

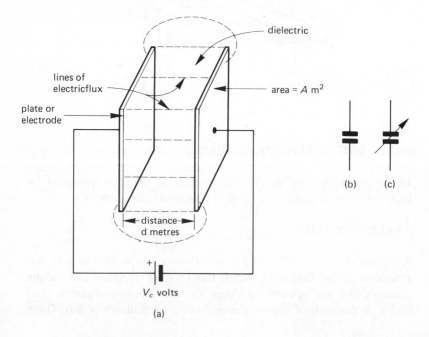

The capacitor in Figure 6.2 is known as a **fixed capacitor** because its geometry is 'fixed', that is, the area of the plates and the distance between the plates cannot be altered. If it is possible to vary either the area of the plates, or the distance between them, or the nature of the dielectric, then the capacitor is known as a **variable capacitor**. Circuit symbols for fixed and variable capacitor are shown in diagrams (b) and (c) of Figure 6.2.

One very useful feature of a capacitor is that it can **store electrical energy**. If, for example, the battery were to be disconnected from the capacitor in Figure 6.2, the capacitor would retain its stored charge for a considerable period of time. The reader should note, however, that *the energy is stored in the dielectric material of the capacitor* and not in the plates.

6.5 **POTENTIAL GRADIENT OR ELECTRIC FIELD INTENSITY**

The **potential gradient** or **electric field intensity**, symbol E, in volts per metre in the *dielectric* is given by

$$\text{electric field intensity}, E = \frac{\text{thickness of dielectric} \quad (m)}{\text{p.d. across dielectric} \quad (V)}$$

In the case of Figure 6.2 this is given by the equation

$$E = \frac{V_C}{d} \text{ V/m}$$

Even in low voltage capacitors the electric field intensity can be very high.

Example
Calculate the electric field intensity between the plates of a capacitor whose dielectric is 0.01 mm thick and which has a p.d. of 10 V between its plates.

Solution
$$d = 0.01 \text{ mm} = 0.01 \times 10^{-3} \text{ m}; V_C = 10 \text{ } V$$

$$\text{Electric field intensity}, E = \frac{V_C}{d} = \frac{10 \text{ } V}{(0.01 \times 10^{-3})}$$

$$= 1\,000\,000 \text{ V/m (Ans.)}$$

6.6 **ELECTROSTATIC SCREENING**

Many items of electrical plant are sensitive to strong electric fields. To protect the apparatus, it is necessary to enclose it within a conducting mesh or sheath as shown in Figure 6.3. A special cage made from wire mesh designed to protect not only apparatus but also humans from intense electric fields is known as a **Faraday cage**. In effect, the mesh has a very low electrical resistance and prevents the electric field from penetrating it.

fig 6.3   *electrostatic screening*

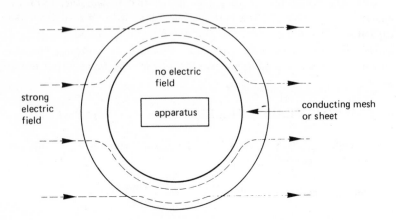

6.7 UNITS OF CAPACITANCE

The **capacitance**, symbol C, of a capacitor is a measure of the ability of the capacitor to store electric charge; the basic unit of capacitance is the **farad** (unit symbol F). Unfortunately the farad is an impractically large unit, and the following submultiples are commonly used

$$1 \text{ microfarad} = 1 \ \mu F = 1 \times 10^{-6} \text{ F} = 0.000\,001 \text{ F}$$
$$1 \text{ nanofarad} = 1 \text{ nF} = 1 \times 10^{-9} \text{ F} = 0.000\,000\,001 \text{ F}$$
$$1 \text{ picofarad} = 1 \text{ pF} = 1 \times 10^{-12} \text{ F} = 0.000\,000\,000\,001 \text{ F}$$

6.8 CHARGE STORED BY A CAPACITOR

Experiments with capacitors show that the electric charge, Q coulombs, stored by a capacitor of capacitance C farads is given by

stored charge, $Q = CV_C$ coulombs

where V_C is the voltage across the capacitor.

Example
Calculate the charge stored by a capacitor of capacitance (i) 10 μF, (ii) 100 pF having a voltage of 10 V between its terminals.

Solution

    (i) $C = 10 \ \mu F = 10 \times 10^{-6}$ F

    $Q = CV_C = (10 \times 10^{-6}) \times 10 = 100 \times 10^{-6}$ C

            $= 100 \ \mu C$ (Ans.)

    (ii) $C = 100$ pF $= 100 \times 10^{-12}$ F

    $Q = CV_C = (100 \times 10^{-12}) \times 10 = 1000 \times 10^{-12}$ C

            $= 1000$ pC or 1 nC

6.9 ENERGY STORED IN A CAPACITOR

During the time that the electric field is being established in the dielectric, energy is being stored. This energy is available for release at a later time when the electric field is reduced in value. The equation for the energy, W joules, stored in a capacitor of capacitance C farads which is charged to V_C volts is

$$\text{energy stored, } W = \frac{CV_C^2}{2} \text{ joules (J)}$$

Example
Calculate the energy stored by a capacitor of $10 \ \mu F$ capacitance which is charged to 15 V.

Solution

    $C = 10 \ \mu F = 10 \times 10^{-6}$ F; $V_C = 15$ V

    Energy stored, $W = \dfrac{CV_C^2}{2} = 10 \times 10^{-6} \times \dfrac{15^2}{2}$

                $= 1.25 \times 10^{-3}$ J or 1.125 mJ (Ans.)

6.10 ELECTRIC FLUX DENSITY

Even though electric flux is an imaginary concept, it has some validity in that its effects can be measured. The electric flux emanates from an electrically charged body, and we can determine the **electric flux density**, symbol D, which is the amount of electric flux passing through a unit area, that is, it is the amount of electric flux passing through an area of $1 \ m^2$.

If, for example, Q lines of electric flux pass through a dielectric of area A m^2, the electric flux density in the dielectric is

$$D = \frac{Q}{A} \text{ coulombs per metre}^2 \text{ (C/m}^2)$$

6.11 PERMITTIVITY OF A DIELECTRIC

The **permittivity**, symbol ϵ, of a dielectric is a measure of the ability of the dielectric to concentrate electric flux. If, for example, a parallel-plate capacitor with air as its dielectric has a flux density in its dielectric of Y C/m^2 then, when the air is replaced by (say) a mica dielectric, the flux density in the dielectric is found to increase to a value in the range $3Y$-$7Y$ C/m^2. That is to say, the mica dielectric produces a higher concentration of electric flux than does the air.

The net result is that a capacitor with a mica dielectric can store more energy than can an air dielectric capacitor. Alternatively, for the same energy storage capacitory, the mica dielectric capacitor is physically smaller than the air dielectric capacitor.

The basic dielectric with which all comparisons are made is a vacuum but, since dry air has similar features, we can regard air as a 'reference' dielectric. Experiments show that the relationship between the electric flux density D and the electric field intensity E (or potential gradient) in a dielectric is given by

$$D = E\epsilon \text{ coulombs per square metre}$$

where ϵ is the **absolute permittivity** of the dielectric. The permittivity of a vacuum (known as the **permittivity of free space**) is given the special symbol ϵ_O and has the dimensions farads per metre (F/m). Its value is

$$\epsilon_O = 8.85 \times 10^{-12} \text{ (F/m)}$$

To all intents and purposes we may regard this as the *permittivity of dry air*.

When air is replaced as the dielectric by some other insulator, the electric flux density is magnified by a factor known as the **relative permittivity**, ϵ_r. The relative permittivity is simply a dimensionless multiplying factor; for most practical dielectrics its value lies in the range 2-7, but may have a value of several hundred in a few cases. Typical values are given in Table 6.1. The absolute permittivity of a material is given by the expression

ϵ = permittivity of free space × relative permittivity

$$= \epsilon_O \epsilon_r \text{ F/m}$$

Table 6.1 *Relative permittivity of various materials*

Material	Relative permittivity
Air	1.0006
Bakelite	4.5–5.5
Glass	5–10
Mica	3–7
Paper (dry)	2–2.5
Rubber	2–3.5

6.12 CAPACITANCE OF A PARALLEL-PLATE CAPACITOR

The potential gradient or electric field intensity, E V/m, in the dielectric of the parallel-plate capacitor in Figure 6.4 is given by

$$E = \frac{V_C}{d} \text{ V/m}$$

where V_C is the p.d. between the plates and d is the thickness of the dielectric. If the capacitor stores Q coulombs of electricity, then the electric flux density, D coulombs per square metre, is

$$D = \frac{Q}{A} \text{ C/m}^2$$

fig 6.4 *capacitance of a parallel-plate capacitor*

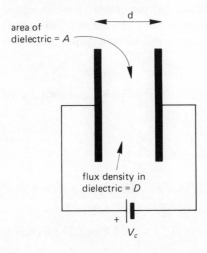

area of dielectric = A

flux density in dielectric = D

V_c

where A is the area of the dielectric (which is equivalent to the area of the plate). From section 6.11, the relationship between E and D in the dielectric is

$$D = E\epsilon \ \text{C/m}^2$$

where ϵ is the absolute permittivity ($\epsilon = \epsilon_0\epsilon_r$) of the dielectric. Hence

$$\frac{Q}{A} = \frac{V_C\epsilon}{d}$$

But the charge stored, Q coulombs, is given by $Q = CV_C$, where C is the **capacitance** of the capacitor (in farads). Therefore

$$\frac{CV_C}{A} = \frac{V_C\epsilon}{d}$$

Cancelling V_C on both sides and multiplying both sides of the equation by A gives the following expression for the capacitance, C, of the parallel-plate capacitor:

$$C = \frac{A\epsilon}{d} = \frac{A\epsilon_0\epsilon_r}{d} \ \text{farads (F)}$$

where A is in m^2, ϵ is in F/m, and d is in m.

Example

A parallel-plate capacitor has plates of area 600 cm^2 which are separated by a dielectric of thickness 1.2 mm whose relative permittivity is 5. The voltage between the plates is 100 V. Determine (a) the capacitance of the capacitor in picofarads, (b) the charge stored in microcoulombs, (c) the electric field intensity in the dielectric in kV/m and (d) the electric flux density in the dielectric in microcoulombs per square metre.

Solution

$A = 600 \ \text{cm}^2 = 600 \times (10^{-2})^2 \ \text{m}^2; d = 1.2 \ \text{mm} = 1.2 \times 10^{-3} \ \text{m};$
$\epsilon_r = 5; V_C = 100$

(a) Capacitance, $C = \epsilon_0\epsilon_r \dfrac{A}{d}$

$$= \frac{(8.85 \times 10^{-12}) \times 5 \times (600 \times 10^{-4})}{1.5 \times 10^{-3}}$$

$$= 1.77 \times 10^{-9} \ \text{F or 1.77 pF (Ans.)}$$

(b)   Charge stored, $Q = CV_C = (1.77 \times 10^{-9}) \times 100$

$$= 1.77 \times 10^{-7} \text{ C}$$

$$= 0.177 \times 10^{-6} \text{ C or } 0.177 \text{ } \mu\text{C (Ans.)}$$

Note

The reader should carefully note the difference here between the use of the symbol C (in italics) to represent the capacitance of the capacitor, and the symbol C (in roman type) to represent the unit of charge (the coulomb).

(c)   Electric field intensity $= \dfrac{V_C}{d} = \dfrac{100}{1.2 \times 10^{-3}}$

$$= 83\,333 \text{ V/m or } 83.333 \text{ kV/m (Ans.)}$$

(d)   Electric flux density, $D = \dfrac{Q}{A}$

$$= \dfrac{1.77 \times 10^{-7}}{(600 \times 10^{-4})}$$

$$= 2.95 \times 10^{-6} \text{ C/m}^2$$

$$\text{or } 2.95 \text{ } \mu\text{C/m}^2 \text{ (Ans.)}$$

6.13 APPLICATIONS OF CAPACITORS

One of the most obvious applications of capacitors is as energy storage elements. However, any capacitor which can store a reasonable amount of energy would be very large indeed and, if only for this reason, it has a very limited use for this type of application.

An application of the capacitor in a number of electrical and electronic circuits is as a 'd.c. blocking' device, which prevents direct current from flowing from one circuit to another circuit. It can be used in this application because the dielectric is an insulator which, once the capacitor is fully 'charged', prevents further direct current from flowing through it.

Capacitors are also widely used in alternating current circuits (see Chapters 11 and 12), and discussion on these applications is dealt with in the specialised chapters.

They are also used in voltage 'multiplying' circuits, in which a voltage much higher than the supply voltage is produced using special circuits using capacitors. An everyday example of this is the electronic flashgun used with a camera; in this case an electronic circuit 'pumps up' the voltage across a 'string' of capacitors, the stored energy being finally discharged into the flash lamp. In industry, capacitors are used to produce voltages of several million volts by this means; the resulting voltage is used to test high voltage apparatus.

Yet another application is the use of a capacitor as a **transducer** or **sensor**. A transducer or sensor is a device which converts energy of one kind to energy of another kind; it can be used, for example, to translate linear movement into a change in voltage. That is, a transducer can be used to *sense a change* in some characteristic. The equation for the capacitance of a parallel-plate capacitor gives us an indication of the way in which the capacitor can be used as a sensor. The equation is

$$C = \frac{\epsilon A}{d}$$

From this equation, the capacitance of the capacitor is

1. proportional to the permittivity ϵ of the dielectric. Any change in the permittivity produces a proportional change in C;
2. proportional to the area A of the dielectric. Any change in this area gives rise to a proportional change in C;
3. inversely proportional to the displacement d between the plates. An increase in d reduces the capacitance, and a reduction in d increases the capacitance.

For example, you can detect the movement of, say, the end of a shaft or beam by mechanically linking the shaft or beam so that it changes one of the factors ϵ, A or d. Several basic techniques are shown in Fiure 6.5.

In Figure 6.5(a), the dielectric material is moved either further into or out of the capacitor to change the average permittivity of the capacitor. In Figure 6.5(b), one plate is moved relative to the other plate to alter the total area of the capacitor. In Figure 6.5(c), one plate is moved either towards or away from the other plate to change d. The net effect of any one of these changes is to change the capacitance of the capacitor, the change in capacitance being related to the displacement.

The change in capacitance produced by any of these methods is usually only a few tens of picofarads but, none the less, the change can be measured electrically.

A capacitor can also be used, for example, to measure fluid pressure or flow simply by allowing the pressure produced by a set of bellows to displace a diaphragm. The movement of the diaphragm can be linked to a capacitor to produce one of the changes in the capacitor illustrated in Figure 6.5. The resulting change in capacitance is related in some way to the pressure or flow being measured.

fig 6.5 *the parallel-plate capacitor as a transducer*

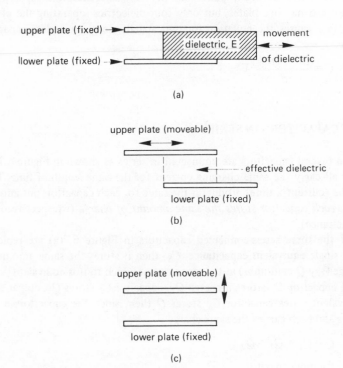

upper plate (fixed) →

dielectric, E

movement

llower plate (fixed) →

of dielectric

(a)

upper plate (moveable)

- - - effective dielectric

lower plate (fixed)

(b)

upper plate (moveable)

lower plate (fixed)

(c)

6.14 MULTI-PLATE CAPACITORS

One method of increasing the capacitance of a capacitor is to increase the area of the dielectric (remember – the energy is stored in the dielectric in a capacitor). A popular method of doing this is to use a multi-plate capacitor of the type shown in Figure 6.6.

fig 6.6 *a multiple-plate capacitor*

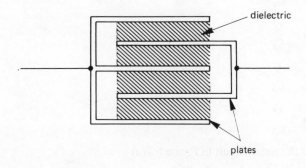

dielectric

plates

A capacitor having n plates has $(n-1)$ dielectrics; the capacitor in Figure 6.6 has five plates, but only four dielectrics separating the plates. This fact causes the equation for the capacitance of an n-plate capacitor to be

$$C = \frac{(n-1)\epsilon A}{d} \ \text{F}$$

6.15 CAPACITORS IN SERIES

When several capacitors are connected in series as shown in Figure 6.7(a), they all carry the same charging current for the same length of time. That is, the (current × time) product is the same for each capacitor; put another way, *each capacitor stores the same amount of charge* (irrespective of its capacitance).

If the three series-connected capacitors in Figure 6.7(a) are replaced by a single **equivalent capacitance**, C_S, then it stores the same amount of charge (say Q coulombs) as the series circuit which to it is equivalent.

If capacitor C_1 stores a charge Q_1, capacitor C_2 stores Q_2, etc, and the equivalent series capacitor C_S stores Q then, since the capacitors are in series and each carries the same charge

$$Q = Q_1 = Q_2 = Q_3$$

Now, for any capacitor

stored charge = capacitance × voltage across the capacitor

then the charge stored by C_1 is $C_1 V_1$, the charge stored by C_2 is $C_2 V_2$, etc, and the charge stored by C_S is $C_S V_S$, hence

$$Q = C_1 V_1 = C_2 V_2 = C_3 V_3 = C_S V_S$$

that is

$$V_1 = \frac{Q}{C_1}$$

$$V_2 = \frac{Q}{C_2}$$

$$V_3 = \frac{Q}{C_3}$$

$$V_S = \frac{Q}{C_S}$$

Now, for the series circuit in Figure 6.7(a)

fig 6.7 *(a) capacitors in series and (b) their electrical equivalent capacitance*

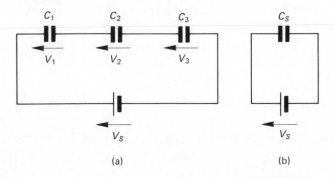

(a)

(b)

$$V_S = V_1 + V_2 + V_3$$

$$\frac{Q}{C_S} = \frac{Q}{C_1} + \frac{Q}{C_2} + \frac{Q}{C_3} = Q\left(\frac{1}{C_1} + \frac{1}{C_2} + \frac{1}{C_3}\right)$$

Cancelling Q on both sides of the above equation gives

$$\frac{1}{C_S} = \frac{1}{C_1} + \frac{1}{C_2} + \frac{1}{C_3}$$

That is, *the reciprocal of the equivalent capacitance of a series connected group of capacitors is the sum of the reciprocals of their respective capacitances.*

It is of interest to note that the *equivalent capacitance, C_S, of the series circuit is less than the smallest individual capacitance in the circuit* (see the example below).

The special case of two capacitors in series
In this case it can be shown that the equation for the equivalent capacitance of the circuit is

$$C_S = \frac{C_1 C_2}{C_1 + C_2}$$

Example
Calculate the equivalent capacitance of three series-connected capacitors having capacitances of 10, 20 and 40 microfarads, respectively.

Solution

$$C_1 = 10\ \mu F;\ C_2 = 20\ \mu F;\ C_3 = 40\ \mu F$$

$$\frac{1}{C_S} = \frac{1}{C_1} + \frac{1}{C_2} + \frac{1}{C_3} = \frac{1}{10} + \frac{1}{20} + \frac{1}{40} \ \mu F^{-1}$$

$$= 0.1 + 0.05 + 0.025 = 0.175 \ \mu F^{-1}$$

hence

$$C_S = \frac{1}{0.175} = 5.714 \ \mu F \ (Ans.)$$

Note

The value of C_S is **less than** the smallest capacitance ($10 \ \mu F$) in the circuit.

6.16 CAPACITORS IN PARALLEL

When capacitors are in parallel with one another (Figure 6.8(a)), they have the same voltage across them. The charge stored by the capacitors in the

fig 6.8 *(a) capacitors in parallel and (b) their electrical equivalent capacitance*

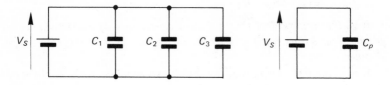

figure is, therefore, as follows

$$Q_1 = C_1 V_S$$

$$Q_2 = C_1 V_S$$

$$Q_3 = C_3 V_S$$

The total charge stored by the circuit is

$$Q_1 + Q_2 + Q_3 = C_1 V_S + C_2 V_S + C_3 V_S = V_S(C_1 + C_2 + C_3)$$

The parallel bank of capacitors in diagram (a) can be replaced by the single equivalent capacitor C_P in diagram (b). Since the supply voltage V_S is applied to C_P, the charge Q_P stored by the equivalent capacitance is

$$Q_P = V_S C_P$$

For the capacitor in diagram (b) to be equivalent to the parallel combination in diagram (a), both circuits must store the same charge when connected to V_S. That is

$$Q_P = Q_1 + Q_2 + Q_3$$

or

$$V_S C_P = V_S(C_1 + C_2 + C_3)$$

when V_S is cancelled on both sides of the equation above, the expression for the capacitance C_P is

$$C_P = C_1 + C_2 + C_3$$

The equivalent capacitance of a parallel connected bank of capacitors is equal to the sum of the capacitances of the individual capacitors. It is of interest to note that *the equivalent capacitance of a parallel connected bank of capacitors is greater than the largest value of capacitance in the parallel circuit.*

Example
Calculate the equivalent capacitance of three parallel-connected capacitors of capacitance 10, 20 and 40 microfarads, respectively.

Solution

$$C_1 = 10 \ \mu F; \ C_2 = 20 \ \mu F; \ C_3 = 40 \ \mu F$$

equivalent capacitance, $C_P = C_1 + C_2 + C_3 = 10 + 20 + 40$

$$= 70 \ \mu F \ (Ans.)$$

Note
The value of C_P is **greater** than the largest capacitance ($40 \ \mu F$) in the circuit.

6.17 CAPACITOR CHARGING CURRENT

In this section of the book we will investigate what happens to the current in a capacitor which is being charged and what happens to the voltage across the capacitor.

The basic circuit is shown in Figure 6.9. Initially, the blade of switch S is connected to contact B, so that the capacitor is discharged; that is the voltage V_C across the capacitor is zero.

You will observe that we now use the *lower-case* letter v rather than the upper case letter V to describe the voltage across the capacitor. The reason

fig 6.9 *charging a capacitor. The blade of switch S is changed from position B to position A at time t = 0*

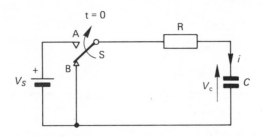

is as follows. Capital letters are used to describe either d.c. values or 'effective a.c.' values (see chapter 10 for details of the meaning of the latter phrase) in a circuit. Lower case (small) letters are used to describe **instantaneous values**, that is, values which may change with time. In this case the capacitor is initially discharged so that at 'zero' time, that is, $t = 0$, we can say that $v_C = 0$. As you will see below, a little time after switch S is closed the capacitor will be, say, half fully-charged (that is $v_C = \frac{V_S}{2}$). As time progresses, the voltage across C rises further. Thus, v_C changes with time and has a different value at each instant of time. Similarly, we will see that the charging current, i, also varies in value with time. We will now return to the description of the operation of the circuit.

When the contact of switch S is changed to position A at time $t = 0$, current begins to flow into the capacitor. Since the voltage across the capacitor is zero at this point in time, the **initial value** of the **charging current** is

$$i = \frac{\text{supply voltage} - \text{voltage across the capacitor}}{\text{circuit resistance}}$$

$$= \frac{(V_S - 0)}{R} = \frac{V_S}{R}$$

Let us call this value i_0 since it is the current at 'zero time'. As the current flows into the capacitor, it begins to acquire electric charge and the voltage across it builds up in the manner shown in Figure 6.10(a).

Just after the switch is closed and for a time less than $5T$ (see Figure 6.10), the current through the circuit and the voltage across each element in the circuit change. This period of time is known as the **transient period** of operation of the circuit. During the transient period of time, the voltage v_C across the capacitor is given by the mathematical expression

$$v_C = V_S (1 - e^{-t/T}) \text{ volts}$$

fig 6.10 *capacitor charging curves for (a) capacitor voltage, (b) capacitor current*

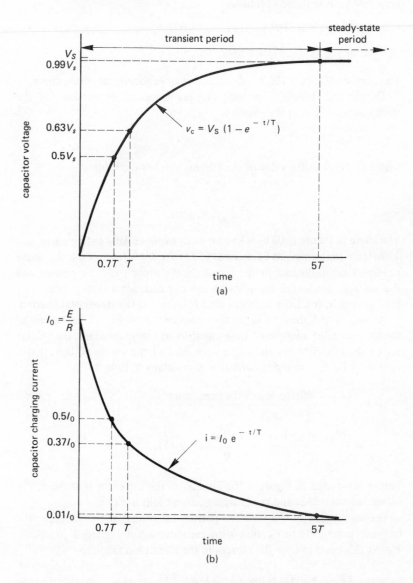

where v_C is the voltage across the capacitor at time t seconds after the switch has been closed, V_S is the supply voltage, T is the *time constant* of the circuit (see section 6.18 for details), and e is the number 2.71828 which is the base of the natural logarithmic series.

For example, if the time constant of an *RC* circuit is 8 seconds, the voltage across the capacitor 10 seconds after the supply of 10 V has been connected is calculated as follows

$$v_C = 10(1 - e^{-10/8}) = 10(1 - e^{-1.125})$$

$$= 10(1 - 0.287) = 7.13 \text{ V}$$

The curve in Figure 6.10(a) is described as an **exponentially rising curve**.

During the transient period, the mathematical expression for the transient current, i, in the circuit is

$$i = I_O e^{-t/T}$$

where I_O is the initial value of the current and has the value

$$I_O = \frac{V_S}{R} \text{ A}$$

The curve in Figure 6.10(b) is known as an **exponentially falling curve**.

After a time equal to $5T$ seconds ($5T = 5 \times 8 = 40$ seconds in the above example) the transients in the circuit 'settle down', and the current and the voltages across the elements in the circuit reach a steady value. The time period beyond the transient time is known as the **steady-state period**.

As mentioned above, T is the *time constant* of the circuit, and it can be shown that after a length of time equal to one time constant, the voltage across the capacitor has risen to 63 per cent of the supply voltage, that is $v_C = 0.63 \ V_S$. The charging current at this instant of time is

$$i = \frac{(V_S - \text{voltage across the capacitor})}{R}$$

$$= \frac{(V_S - 0.63 \ V_S)}{R} = \frac{0.37 \ V_S}{R} = 0.37 I_O$$

This is illustrated in Figure 6.10. That is, as the capacitor is charged, the voltage across it rises and the charging current falls in value.

On completion of the transient period, the voltage across the capacitor has risen practically to V_S, that is the capacitor is 'fully charged' to voltage V_S. At this point in time the current in the circuit has fallen to

$$i = \frac{(V_S - \text{voltage across } C)}{R} \simeq \frac{(V_S - V_S)}{R} = 0$$

Thus, when the capacitor is fully charged, **it no longer draws current from the supply**.

6.18 **THE TIME CONSTANT OF AN** RC **CIRCUIT**

For a circuit containing a resistor R and a capacitor C, the **time constant**, T, is calculated from

$T = RC$ seconds

where R is in ohms and C is in farads. For example, if R = 2000 Ω and C = 10 μF, then

$T = RC = 2000 \times (10 \times 10^{-6}) = 0.02$ s or 20 ms

If the supply voltage is 10 V, it takes 0.02 s for the capacitor to charge to 0.63 VS = 0.63 × 10 = 6.3 V.

The time taken for the transients in the circuit to vanish and for the circuit to settle to its steady-state condition (see Figure 6.10 is about $5T$ seconds. This length of time is referred to as the **settling time**. In the above case, the settling time is 5 × 0.02 = 0.1 s.

6.19 **CAPACITOR DISCHARGE**

While the contact of switch S in Figure 6.11 is in position A, the capacitor is charged by the cell. When the contact S is changed from A to B, the capacitor is discharged via resistor R.

Whilst the capacitor discharges current through resistor R, energy is extracted from the capacitor so that the voltage v_C across the capacitor gradually decays towards zero value. When discharging energy, current flows out of the positive plate (the upper plate in Figure 6.11); that is, the current in Figure 6.11 flows in the reverse direction when compared with the charging condition (Figure 6.9).

fig **6.11** *capacitor discharge. The blade of switch S is changed from A to B at time t = 0*

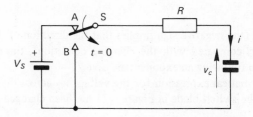

The graph in Figure 6.12(a) shows how the capacitor voltage decays with time. The graph in diagram (b) shows how the discharge current rises to a maximum value of $\dfrac{-V_S}{R}$ at the instant that the switch blade is moved

fig 6.12 *capacitor discharge curves for (a) capacitor voltage, (b) capacitor current*

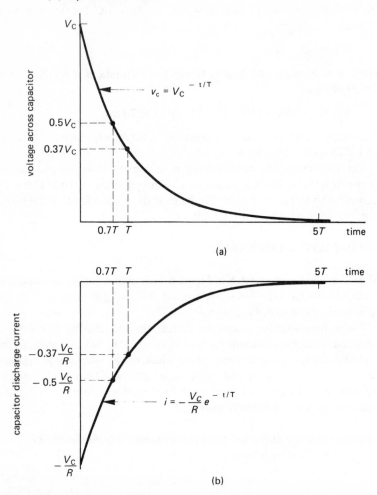

(a)

(b)

to position B (the negative sign implies that the direction of the current is reversed when compared with the charging condition); the current then decays to zero following an exponential curve.

The mathematical expression for the voltage v_C across the capacitor at time t *after* the switch blade in Figure 6.11 has been changed from A to B is

$$v_C = V_C e^{t/T} \text{ volts}$$

where V_C is the voltage to which the capacitor has been charged just before the instant that the switch blade is changed to position B. The time constant of the circuit is $T = RC$ (T in seconds, R in ohms, C in farads), and $e = 2.71828$. The expression for the discharge current is

$$i = - \frac{V_C}{R} e^{-t/T}$$

Once again, it takes approximately $5T$ seconds for the transient period of the discharge to decay, during which time the current in the circuit and the voltage across the circuit elements change. When the steady-stage period is reached, the current in the circuit and the voltage across R and C reach a steady value (zero in this case, since the capacitor has discharged its energy).

Theoretically, it takes an infinite time for the transient period to disappear but, in practice, it can be thought of as vanishing in a time of $5T$.

6.20 TYPES OF CAPACITOR

Capacitors are generally classified according to their dielectrics, for example, paper, polystyrene, mica, etc. The capacitance of all practical capacitors varies with age, operating temperature, etc, and the value quoted by the manufacturer usually only applies under specific operating conditions.

Air dielectric capacitors Fixed capacitors with air dielectrics are mainly used as laboratory standards of capacitance. Variable capacitance air capacitors have a set of fixed plates and a set of moveable plates, so that the capacitance of the capacitor is altered as the overlapping area of the plates is altered.

Paper dielectric capacitors In one form of paper capacitor, shown in Figure 6.13, the electrodes are metal foils interleaved with layers of paper which have been impregnated with oil or wax with a plastic (polymerisable) impregnant. In the form of construction shown, contact is made between the capacitor plates and the external circuit via pressure contacts.

In capacitors known as **metallised paper capacitors**, the paper is metallised so that gaps or voids between the plates and the dielectric are avoided. Important characteristics of this type when compared with other 'paper' types are their small size and their 'self-healing' action after electrical breakdown of the dielectric. In the event of the paper being punctured when a transient voltage 'spike' is applied to the terminals of the capacitor (this is a practical hazard for any capacitor), the metallising in the region of the puncture rapidly evaporates and prevents the capacitor from developing a short-circuit.

Plastic film dielectric capacitors These use plastic rather than paper dielectric and are widely used. The production techniques provide low

fig 6.13 *construction of one form of tubular paper capacitor*

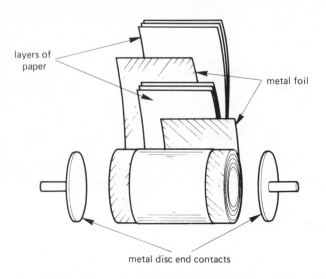

layers of paper

metal foil

metal disc end contacts

cost, high reliability capacitors. The construction is generally as for paper capacitors, typical dielectrics being polystyrene, polyester, polycarbonate and polypropylene.

Mixed dielectric capacitors Capacitors with dielectrics incorporating plastic films and impregnated paper permit the manufacture of low volume, high voltage capacitors.

Mica dielectric capacitors Mica is a mineral which can readily be split down into thin uniform sheets of thickness in the range 0.025 mm–0.075 mm. In the **stacked construction** (see Figure 6.6) mica and metal foil are interleaved in the form of a multiple plate capacitor, the whole being clamped together to maintain a rigid structure. As with paper capacitors, voids between the foil and the dielectric can be eliminated by metallising one side of the mica (**silvered mica capacitors**).

Ceramic dielectric capacitors These capacitors consist of metallic coatings (usually silver) on opposite faces of discs, cups and tubes of ceramic material. The construction of one form of tubular ceramic capacitor is shown in Figure 6.14, the sectional view through the right-hand end illustrating how the connection to the inner electrode is made. Ceramic capacitors are broadly divided into two classes, namely those with a low value of relative permittivity (**low-ϵ** or **low-k** types), having permittivities in the range 6–100, and those with high values of permittivity (**high-ϵ** or **high-k** types), having permittivities in the range 1500–3000.

Low-ϵ types have good capacitance stability and are used in the tuning circuits of electronic oscillators to maintain the frequency of oscillation

fig 6.14 *cut-away section of a tubular ceramic capacitor*

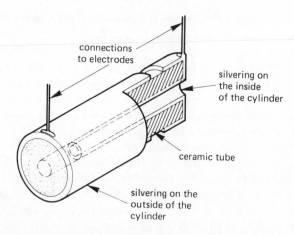

within close limits. High-ϵ types provide a greater capacitance per unit volume than do low-ϵ types, but are subject to a wider range of capacitance variation. They are used in a wide range of electronic applications.

Electrolytic capacitors The dielectric in these capacitors is a thin film of oxide formed either on one plate or on both plates of the capacitor, the thickness of the film being only a few millionths of a centimetre. A consequence is that electrolytic capacitors not only have the highest capacitance per unit volume of all types of capacitor but are the cheapest capacitor per unit capacitance. Offset against these advantages, electrolytic capacitors have a relatively high value of leakage current (particularly so in aluminium electrolytic capacitors) and have a wide variation in capacitance value (from −20 per cent to +50 or +100 per cent in some types).

The majority of electrolytic capacitors are **polarised**, that is the p.d. between the terminals *must* have the correct polarity, otherwise there may be risk of damage to the capacitor. A typical circuit symbol used to represent an electrolytic capacitor is shown in Figure 6.15; the symbol indicates the correct polarity to which the capacitor must be connected.

Although many metals can be coated with an oxide film, it is found that aluminium and tantalum exhibit the best features for use in electro-

fig 6.15 *a circuit symbol for a polarised electrolytic capacitor*

lytic capacitors. The basis of capacitors using the two materials is described below.

After long periods of inactivity, for instance, if they have been stored for several months, the electrodes of electrolytic capacitors require 'reforming' (this applies particularly to aluminium electrolytics). This is carried out by applying the full rated voltage via a 10 kΩ resistor until the leakage current has fallen to its rated value. Should this not be done, there is a risk when the full voltage is applied that the initial value of the leakage current may be large enough to generate an excessive gas pressure inside the capacitor with consequent hazard of an explosion.

Aluminium electrolytic capacitors The basic construction of a polarised aluminium electrolytic capacitor is shown in Figure 6.16. The anode (positive) foil has an oxide film formed on its surface, the latter being the dielectric with a relative permittivity of about 7–10. The cathode (negative) foil is in contact with the actual cathode electrode, which is a paper film impregnated in an electrolyte such as ammonium borate. The physical construction of tubular capacitors is generally similar to that in Figure 6.13, with the impregnated paper and the aluminium foils rolled in the form of a cylinder.

Non-polar electrolytic capacitors, suitable for use either with d.c. or a.c. supplies, are manufactured by forming oxide layers on both foils.

A feature of electrolytic capacitors is that, at high frequency, they 'appear' to the external circuit as inductors. This effect can be overcome by connecting, say, a polycarbonate capacitor of low value in parallel with the electrolytic capacitor.

Tantalum electrolytic capacitors Two types of tantalum capacitor are available, and are those which use foil electrodes and those which employ

fig 6.16 *basic construction of a polarised electrolytic capacitor*

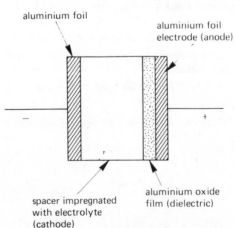

aluminium foil

aluminium foil
electrode (anode)

spacer impregnated
with electrolyte
(cathode)

aluminium oxide
film (dielectric)

a tantalum slug as the anode electrode. The construction of tantalum foil types is similar to that of aluminium foil types.

Tantalum capacitors, although more costly per microfarad than aluminium electrolytics, are more reliable and are physically smaller than their aluminium counterparts. The dielectric properties of tantalum oxide are generally superior to those of aluminium oxide, and result in lower leakage current, a longer 'shelf' life, and a flatter temperature–capacitance curve than aluminium capacitors.

SELF-TEST QUESTIONS

1. What is meant by (i) a line of electric flux, (ii) the direction of electric flux and (iii) electric flux density?
2. A dielectric has an electric field intensity in it of 10 MV/m. If the dielectric thickness is 0.1 mm, what p.d. is applied across it?
3. Explain the principle of electrostatic screening. What is a Faraday cage and how is it used?
4. A capacitor stores a charge of 10 μC. Calculate the p.d. between the terminals of the capacitor if its capacitance is (i) 1.0 μF, (ii) 100 pF.
5. What is meant by the permittivity of a dielectric? How does the permittivity of the dielectric affect the capacitance of a parallel-plate capacitor?
6. Calculate the capacitance of a three-plate capacitor if the area of each plate is 500 cm^2, the plates being separated by a dielectric of thickness 1 mm whose relative permittivity is 4.
7. Capacitors of 2, 4 and 6 μF, respectively, are connected (i) in parallel, (ii) in series. Calculate the resultant capacitance in each case.
8. If each of the capacitor combinations in question 6 has a voltage of 10 V across it, calculate in each case the stored charge and the stored energy.
9. An R–C series circuit contains a 1 kΩ resistor and a 10 μF capacitor, the supply voltage being 10 V d.c. Determine (i) the time constant of the circuit, (ii) the initial value of the charging current, (iii) the time taken for the current to have fallen to 0.005 A, (iv) the 'settling' time of the transients in the circuit and (v) the final value of the capacitor voltage and its charging current.

SUMMARY OF IMPORTANT FACTS

Like electrostatic charges **repel** one another and **unlike** charges **attract** one another.

A **parallel-plate capacitor** consists of two parallel **plates** separated by a **dielectric**. A capacitor can **store electrical energy**.

Apparatus can be **screened** from an electrical field by a wire-mesh **Faraday cage**.

The **charge (Q) stored** by a capacitor is given by $Q = CV$ coulombs.

Permittivity is a measure of the property of a dielectric to **concentrate electric flux**. The **absolute permittivity**, ϵ is given by the product $\epsilon_O \times \epsilon_r$ where ϵ_O is the **permittivity free space** and ϵ_r is the **relative permittivity** (a dimensionless number).

For **series-connected capacitors**, the *reciprocal of the equivalent capacitance is the sum of the reciprocals of the individual capacitances*. The equivalent capacitance of the series circuit is **less than** the smallest value of capacitance in the circuit.

For **parallel-connected capacitors**, the *equivalent capacitance is the sum of the individual capacitances*. The equivalent capacitance is **greater than** the largest value of capacitance in the circuit.

The **initial value of the charging current** drawn by a *discharged capacitor* is $\frac{V_S}{R}$, where V_S is the supply voltage and R is the circuit resistance. When the capacitor is *fully charged*, the charging current is zero. The **time constant**, T, of an R–C circuit is RC **seconds** (R in ohms, C in farads), and the **settling time** (the time taken for the transients to have 'settled out') is *approximately 5T seconds*.

The **initial value of the discharge current** from a capacitor charged to voltage V_C is $\frac{V_C}{R}$, where R is the resistance of the discharge path. The **settling time** of the discharge transients is *approximately 5T seconds*.

The **energy stored** by capacitor C which is charged to a voltage V_C is $\frac{CV_C^2}{2}$ joules (C in farads, V_C in volts).

ELECTROMAGNETISM

7.1 MAGNETIC EFFECTS

Purely 'electrical' effects manifest themselves as voltage and current in a circuit. It was discovered early in the nineteenth century that electrical effects also produce magnetic effects, and it is to this that we direct studies in this chapter.

A magnetic 'field' or magnetic 'force' cannot be seen by man but (and very importantly) we can detect its existence by its effects on other things. For example, when a current flows in a wire, it produces a magnetic field which can be used to attract a piece of iron (this is the principle of the electromagnet), that is to say, a mechanical force exists between the two (this is also the general basis of the electric motor).

Alternatively, if a permanent magnet is quickly moved across a conductor, that is, the magnetic field 'cuts' the conductor, an e.m.f. is induced in the conductor (this is the basis of the electrical generator).

The presence of a magnetic field is detected by its effects, such as its effect on iron filings. The *direction* in which a magnetic field acts at a particular point in space is said to be the direction in which the force acts on an *isolated north-seeking pole* (a *N-pole*) at that point [we may similarly say that a *S-pole* is a *south-seeking pole*].

Experiments with a pair of permanent magnets show two important features, namely:

1. **like magnetic poles**, that is, two N-poles or two S-poles, **repel one another**;
2. **unlike magnetic poles attract one another**.

Consequently, if an isolated N-pole is placed in a magnetic field, it is repelled by the N-pole which produced the field and towards the S-pole. If the isolated N-pole was free to move, it would *move away from the N-pole of the magnet producing the field and towards the S-pole*. It is for this reason that we say that **lines of magnetic flux leave the N-pole of a magnet and enter the S-pole**. In fact, nothing really 'moves' in a magnetic field, and the 'movement' of magnetic flux is a man-made imaginary concept (none the less, it is sometimes convenient to refer to the 'direction' of the magnetic flux).

However, before studying electromagnetism (that is, magnetism produced by electrical effects) we must understand the basic mechanics of magnetism itself.

7.2 MAGNETISM

The name **magnetism** is derived from **magnetite**, which is an iron-oxide mineral whose magnetic properties were discovered before man grasped the basic principles of electricity. It was found that a *magnet* had the property of attracting iron and similar materials; the word **ferromagnetism** is frequently used in association with the magnetic properties of iron.

To understand the nature of a magnet we must take a look at the composition of iron. In an atom, the spinning motion of the electrons around the atomic nucleus is equivalent to an electric current in a loop of wire. In turn, this electric current produces a small magnetic effect. There are many electrons associated with each atom, and the magnetic effect of some electrons will cancel others out. In iron there is a slight imbalance in the electron spins, giving rise to an overall magnetic effect. Groups of atoms appear to 'club' together to produce a small permanent magnet, described by scientists and engineers as a **magnetic domain**. Although a magnetic domain contains billions of atoms, it is smaller in size than the point of a needle.

Each domain in a piece of iron has a N-pole and a S-pole but, in a demagnetised piece of iron, the domains point in random directions so that the net magnetic field is zero, as shown in Figure 7.1(a). Each arrow in the figure represents a magnetic domain.

When an external magnetic field is applied to the iron in the direction shown in Figure 7.1(b), some of the magnetic domains turn in the direction of the applied magnetic field and remain in that direction after the field is removed. That is, the external magnetic field causes the piece of iron to be partially magnetised so that it has a N-pole and a S-pole. Repeated application of the external magnetic field results in the iron becoming progressively more magnetised as more of the domains align with the field. A bar of iron can be magnetised in this way simply by stroking it along its length in one direction with one pole of a permanent magnet.

If the magnetic field intensity is increased, all the domains ultimately align with the applied magnetic field as shown in Figure 7.1(a). When this occurs, the iron produces its maximum field strength and the bar is said to

fig 7.1 *magnetising a piece of iron*

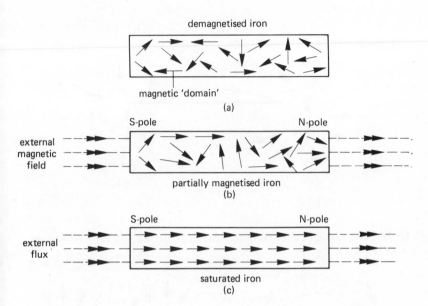

demagnetised iron

magnetic 'domain'

(a)

external magnetic field

S-pole   N-pole

partially magnetised iron

(b)

external flux

S-pole   N-pole

saturated iron

(c)

be magnetically **saturated**. Since no further domains remain to be aligned, any further increase in the external magnetic field does not produce any significant increase in the magnetism of the iron.

7.3 MAGNETIC FIELD PATTERN OF A PERMANENT MAGNET

The magnetic field pattern of a permanent magnet can be traced out by covering the magnet with a piece of paper and sprinkling iron filings on it. When the paper is tapped, the iron filings take up the pattern of the magnetic field (see Figure 7.2). You will note that the flux lines are shown to 'leave' the N-pole and to 'enter' the S-pole.

When unlike magnetic poles are brought close together (see Figure 7.3(a)), the flux leaves the N-pole of one magnet and enters the S-pole of the other magnet, and there is a force of attraction between them. When unlike magnetic poles are brought close together (Figure 7.3(b)) there is a force of repulsion between the poles.

7.4 DIRECTION OF THE MAGNETIC FIELD AROUND A CURRENT-CARRYING CONDUCTOR

Experiments carried out with simple equipment such as a compass needle and a conductor which is carrying current allow us to determine the

fig 7.2   *the flux pattern produced by a permanent magnet*

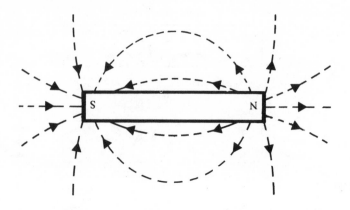

fig 7.3   *the magnetic flux pattern between (a) unlike poles, (b) like poles*

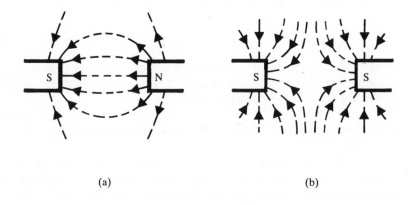

(a)   (b)

'direction' of the magnetic flux around the conductor, as shown in Figure 7.4(a). If the direction of the current in the conductor is reversed, the compass needle reverses direction, indicating that the magnetic field has reversed.

A simple rule known as the **screw rule** allows us to predict the direction of the magnetic field around the conductor as follows:

> **If you imagine a wood screw (Figure 7.4(b)) pointing in the direction of current flow, the direction of the magnetic field around the conductor is given by the direction in which you turn the head of the screw in order to propel it in the direction of the current.**

fig 7.4 *(a) direction of the magnetic flux around a current-carrying conductor, (b) the screw rule for determining the direction of the flux*

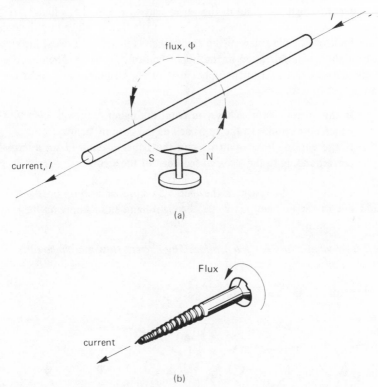

flux, Φ

S   N

current, *I*

(a)

Flux

current

(b)

7.5 SOLENOIDS AND ELECTROMAGNETS

We now turn our attention to the production of magnetism by electrical means. It has been established earlier that whenever current flows in a conductor, a magnetic flux appears around the conductor. If the conductor is wound in the form of a coil, it is known as a **solenoid** if it has an air core and, if it has an iron core, it is known as an **electromagnet**. The current which flows in the coil to produce the magnetic flux is known as the **excitation current**.

The essential difference between an electromagnet and a solenoid is that, for the same value of current, the electromagnet produces a much higher magnetic flux than does the solenoid. This is described in detail in section 7.13.

7.6 FLUX DISTRIBUTION AROUND A CURRENT-CARRYING LOOP OF WIRE

Consider the single-turn coil in the lower half of Figure 7.5, which carries current I. We can determine the direction of the flux produced by the coil if we look on the plan view of the coil (which is the section taken through A–B of the coil), illustrated in the upper half of Figure 7.5. The direction of the current is shown in the upper part of the diagram using an 'arrow' notation as follows:

If the current approaches you from the paper, you will see a 'dot' which corresponds to the approach of the current 'arrow'.
If the current leaves you (enters the paper), you will see a 'cross' corresponding to the 'crossed feathers' of the arrow.

In Figure 7.5, the arrow in the plan view approaches you on the left-hand side of the coil and leaves on the right-hand side. Applying the rule

fig 7.5 *magnetic flux pattern produced by current flow in a single-turn coil*

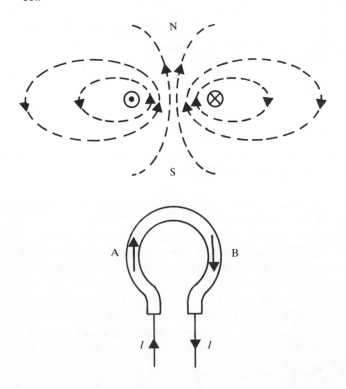

outlined in section 7.4 for the direction of the magnetic flux you will find that, in the plan view, the magnetic flux pattern is as shown.

Referring to the end elevation of the loop (shown in the lower part of Figure 7.5) magnetic flux enters the loop from the side of the page at which you are looking (that is, you are looking at a S-pole) and leaves from the opposite side of the page (that is, the N-pole of the electromagnet is on the opposite side of the loop).

If the direction of the current in the coil is reversed, the magnetic polarity associated with the loop of wire also reverses. You may like to use the information gained so far to verify this.

Simple rules for predicting the magnetic polarity of a current-carrying loop of wire are shown in Figure 7.6.

7.7 MAGNETIC FIELD PRODUCED BY A CURRENT-CARRYING COIL

A coil of wire can be thought of as many single-turn loops connected together. To determine the magnetic polarity of such a coil, we can use the methods outlined in section 7.6.

The first step is to determine the direction of the magnetic flux associated with each wire. This is illustrated in Figure 7.7, and the reader will note that inside the coil the flux produced by the conductors is *additive*. That is the flux produced by one turn of wire reinforces the flux produced by the next turn and so on. In the case shown, the magnetic flux leaves the left-hand side of the coil (the N-pole and enters the right-hand side (the S-pole).

Alternatively, you can deduce the polarity at either end of the coil by 'looking' at the end of the coil and using the method suggested in section 7.6.

If EITHER the direction of the current in the coil is reversed OR the direction in which the coil is wound is reversed (but not both), the magnetic polarity produced by the coil is reversed. The reader is invited to verify this.

fig 7.6 *magnetic polarity of a current-carrying loop of wire*

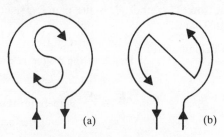

fig 7.7 *magnetic field produced by a coil of wire which carries current*

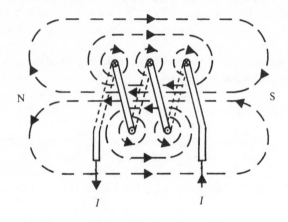

7.8 MAGNETOMOTIVE FORCE OR m.m.f.

In a coil it is the **magnetomotive force**, F, which causes the magnetic flux to be established inside the coil (its analogy in the electric circuit is the e.m.f. which establishes the current in an electric circuit). The equation for m.m.f. is

  m.m.f., F = number of turns × current in the coil

    = NI ampere turns or amperes

The dimensions of m.m.f. are (amperes × turns) or ampere–turns but, since the number of turns is a dimensionless quantity, the dimensions are, strictly speaking, amperes. However, to avoid confusion with the use of amperes to describe the current in the coil, we will use the ampere-turn unit for m.m.f.

7.9 MAGNETIC FIELD INTENSITY OR MAGNETISING FORCE

The **magnetic field intensity** (also known as the **magnetic field strength** or the **magnetising force**), symbol H, is the m.m.f. per unit length of the magnetic circuit. That is:

$$\text{magnetic field intensity} = \frac{\text{m.m.f.}}{\text{length of the magnetic circuit}}$$

$$= \frac{F}{l} = \frac{NI}{l} \text{ ampere-turns per metre}$$
$$\text{or amperes per metre}$$

Once again, to avoid confusion between the use of the ampere and the ampere-turn when used in the electromagnetic context, we will use the ampere-turn per metre unit.

Example

A coil has 1000 turns of wire on it and carries a current of 1.2 A. If the length of the magnetic circuit is 12 cm, calculate the value of the m.m.f. produced by the coil and also the magnetic field intensity in the magnetic circuit.

Solution

$$N = 1000 \text{ turns}; I = 1.2 \text{ A}; l = 12 \text{ cm} = 12 \times 10^{-2} \text{ m}$$

m.m.f., $F = NI = 1000 \times 1.2 = 1200$ ampere-turns (Ans.)

magnetic field intensity, $H = \dfrac{F}{l} = \dfrac{1200}{12 \times 10^{-2}}$

$$= 10\,000 \text{ ampere-turns per metre}$$

7.10 MAGNETIC FLUX

The **magnetic flux**, symbol Φ, produced by a magnet or electromagnet is a measure of the magnetic field, and has the unit of the **weber** (Wb).

7.11 MAGNETIC FLUX DENSITY

The **magnetic flux density**, symbol B is the amount of magnetic flux passing through an area of 1 m^2 which is perpendicular to the direction of the flux; the unit of flux density is the **tesla** (T). That is

$$\text{flux density}, B = \frac{\text{flux}, \Phi}{\text{area}, A} \text{ tesla } (T)$$

where Φ is in Wb and A in m^2. A flux density of 1 T is equivalent to a flux of 1 weber passing through an area of 1 square metre, or

$$1\ T = 1\ Wb/\text{m}^2$$

Example

The magnetic flux in the core of an electromagnet is 10 mWb, the flux density in the core being 1.2 T. If the core has a square cross-section, determine the length of each side of the section.

Solution

$$\Phi = 10 \text{ mWb} = 10 \times 10^{-3} \text{ Wb}; B = 0.75 \text{ T}$$

Since

$$B = \frac{\Phi}{A}$$

then

$$\text{area of core}, A = \frac{\Phi}{B} = \frac{10 \times 10^{-3}}{1.2} = 0.0083333 \text{ m}^2$$

If each side of the section is x metres in length, then

$$x^2 = 0.0083333 \text{ m}^2$$

or

$$x = 0.00913 \text{ m or } 9.13 \text{ cm}.$$

7.12 PERMEABILITY

When an electromagnetic is energised by an electric current, the magnetic field intensity (H ampere turns per metre) produced by the coil establishes a magnetic flux in the magnetic circuit. In turn, this gives rise to a magnetic flux density, B tesla, in the magnetic circuit. The relationship between B and H in the circuit is given by

$$B = \mu H$$

where μ is the **permeability** of the magnetic circuit; the value of the permeability of the magnetic circuit gives an indication of its ability to concentrate magnetic flux within the circuit (the higher its value, the more flux the circuit produces for a particular excitation current).

The **permeability of free space** (or of a vacuum) is given the special symbol μ_O, and is sometimes referred to as the **magnetic space constant**. Its value is

$$\mu_O = 4\pi \times 10^{-7} \text{ henrys per metre (H/m)}$$

where the henry (H) is the unit of inductance and is described in section 7.21. For all practical purposes, *the permeability of air has the same value as the magnetic space constant*.

If a coil has an iron core placed in it, it is found that the magnetic flux produced by the coil increases significantly. The factor by which the flux density increases is given by the **relative permeability**, μ_r, of the material, where

$$\mu_r = \frac{\text{flux density with the iron core}}{\text{flux density with an air (or vacuum) core}}$$

$$= \frac{\mu H}{\mu_O H} = \frac{\mu}{\mu_O}$$

where μ_r is dimensionless. Hence the **absolute permeability**, μ, is given by

$$\mu = \mu_O \mu_r \ \text{H/m}$$

Therefore

$$B = \mu H = \mu_O \mu_r H \ \text{T}$$

The value of μ_r may have a value in the range 1.0–7000, and depends not only on the type of material but also on the operating flux density and temperature.

Example
An electromagnet produces a magnetic field intensity of 500 ampere-turns per metre in its iron circuit. Calculate the flux density in the core if it (i) has an air core, (ii) has a cast-iron core with $\mu_r = 480$, (iii) has a transformer steel core with $\mu_r = 1590$.

Solution

$H = 500$ ampere-turns per metre.

(i) Air core: $\mu = \mu_O = 4\pi \times 10^{-7}$

$B = \mu H = 4\pi \times 10^{-7} \times 500$

$\qquad = 0.628 \times 10^{-3} \ T$ or 0.628 mT (Ans.)

(ii) Cast-iron core: $\mu = \mu_r \mu_O = 480 \times 4\pi \times 10^{-7}$ H/m

$B = \mu H = 480 \times 4\pi \times 10^{-7} \times 500 = 0.302$ T (Ans.)

(iii) Transformer steel core: $\mu = \mu_r \mu_O$

$\qquad\qquad\qquad = 1590 \times 4\pi \times 10^{-7}$ H/m

$B = \mu H = 1590 \times 4\pi \times 10^{-7} \times 500 = 1.0$ T (Ans.)

7.13 MAGNETISATION CURVE FOR IRON AND OTHER FERRO-MAGNETIC MATERIAL

If a coil has an air core, and if the current in the coil (and therefore the magnetic field intensity) is gradually increased from zero, the magnetic flux density in the air core increases in a linear manner as shown by the straight line graph at the bottom of Figure 7.8. The curve relating the flux density B to the magnetic field intensity H is known as the **magnetisation curve** of the material. Since in an air core $B = \mu_O H$, the magnetisation curve is a straight line having a slope of μ_O.

If the air is replaced by a **ferromagnetic material** such as iron, we find that the flux density in the core increases very rapidly between O and A (see Figure 7.8), after which the slope of the curve reduces. Finally, at a very high value of H (corresponding to large value of excitation current), the magnetisation curve for the iron becomes parallel to that for the air core. The reason for the shape of the curve is shown in the next paragraph.

As the excitation current in the coil surrounding the iron increases, the magnetic domains in the iron (see section 7.2) begin to align with the magnetic field produced by the current in the coil. That is, *the magnetism of the domains adds to the magnetic field of the excitation current*; as the current increases, more and more of the domains in the iron align with the external magnetic field, resulting in a rapid increase in the flux density in the iron.

fig 7.8 *magnetisation curves for air and iron*

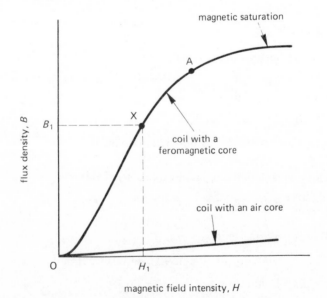

However, when point A in Figure 7.8 is reached, most of the domains have aligned with the field produced by the magnetising current, and beyond this point the rate of increase of the flux density diminishes. This denotes the onset of **magnetic saturation** in the iron.

Any further increase in flux density is caused by the remaining few domains coming into alignment with the applied magnetic field. Finally, the magnetic flux density increases very slowly, and is due only to the increase in excitation current (since all the domains are aligned with the field, no more magnetism can be produced by this means); at this point the magnetisation curve for iron remains parallel to the curve for air.

In a number of cases the magnetisation curve is a useful design tool, allowing calculations on magnetic circuits to be simplified. For example if the value of H is known, the flux density in the iron can be determined directly from the curve. Additionally, the magnetisation curve for a meterial allows us to determine the permeability at any point on the curve; for example, the absolute permeability μ_1 of the iron at point X in Figure 7.8 is

$$\mu_1 = \frac{B_1}{H_1}$$

Example

The following points lie on the magnetisation curve of a specimen of cast steel. Calculate the relative permeability at each point.

Flux density, T	m.m.f., At/m
0.04	100
0.6	600
1.4	2000

Solution

Since $B = \mu_O \mu_r H$ then $\mu_r = \dfrac{B}{\mu_O H}$

For $B = 0.04\ T$, $\mu_r = \dfrac{0.04}{(4\pi \times 10^{-7} \times 100)} = 318$ (Ans.)

For $B = 0.6\ T$, $\mu_r = \dfrac{0.6}{(4\pi \times 10^{-7} \times 600)} = 796$ (Ans.)

For $B = 1.4\ T$, $\mu_r = \dfrac{1.4}{(4\pi \times 10^{-7} \times 2000)} = 557$ (Ans.)

7.14 HYSTERESIS LOOP OF A FERROMAGNETIC MATERIAL

If, after causing the iron core of an electromagnet to be saturated (see the dotted curve in Figure 7.9) the excitation current is reduced, you will find that the B–H curve does not return to the zero when the excitation current is reduced to zero. In fact, the iron retains a certain amount of flux density. This is known as the **remanent flux density**, B_r. The reason is that some of the magnetic domains do not return to their original random direction, and still point in the direction of the original magnetising field.

To reduce the flux density to zero, it is necessary to reverse the direction of the excitation current, that is, to apply a negative value of H. At some reverse magnetising force known as the **coercive force**, H_C, the flux density reaches zero.

If the reverse magnetising current is increased in value, the flux density begins to increase in the reverse direction until, finally, the iron becomes saturated in the reverse direction.

Reducing the magnetising current again to zero leaves a remanent flux density once more but in the reverse direction. The flux density is reduced to zero by increasing the magnetising force in the 'forward' direction; if the 'forward' magnetising current is increased to a high value, the iron becomes saturated once more.

fig 7.9 *hysteresis loop of a ferromagnetic material*

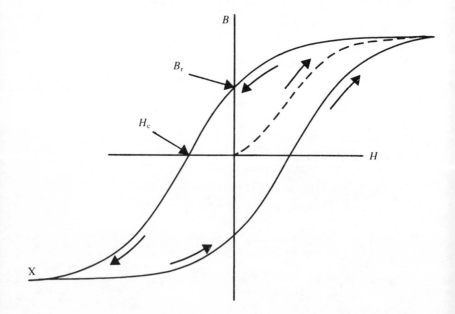

The complete loop in Figure 7.9 is known as the **hysteresis loop** or **B-H loop** of the material. The process of repeatedly reversing the magnetic domains as the B-H loop is cycled (as it is when an alternating current [a.c.] is passed through the coil) results in energy being consumed each time the domains are reversed. The energy consumed in this manner appears in the form of heat in the iron and is known as the **hysteresis energy loss**. The total energy consumed depends on a number of factors including the volume (v) of the iron, the frequency (f) of the magnetic flux reversals, and on the maximum flux density (B_m). The hysteresis power loss, P_n is given by

$$P_h = kvfB_m^n \text{ watts}$$

where k is a number known as the **hysteresis coefficient**, and n is a number known as the **Steinmetz index** and has a value in the range 1.6–2.2.

7.15 'SOFT' AND 'HARD' MAGNETIC MATERIALS

Iron used for electromagnets and relays which must lose their magnetism quickly when the current is switched off, are known as **magnetically soft** materials. They are characterised by a narrow B-H loop with a low remanent flux density and low coercive force (see Figure 7.10(a)).

Iron and steel used for permanent magnets must not only retain a high flux density but also be difficult to demagnetise. These are known as **magnetically hard** materials; they have a high value of B_r (about 1 T) and a high value of H_C (about 50 000 At/m). Their B-H loop is 'square', as shown in Figure 7.10(b). Magnetically hard materials consist of iron combined with small amounts of aluminium, nickel, copper and cobalt (for example Alnico, Alcomax III, Triconal G, etc).

fig 7.10 *B-H curve for (a) a 'soft' magnetic material and (b) a 'hard' magnetic material*

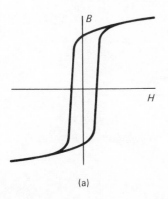
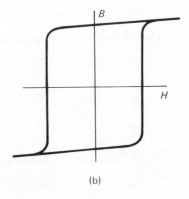

(a)   (b)

7.16 MAGNETIC CIRCUIT RELUCTANCE

The magnetic circuit is an almost exact analogy of the electric circuit insofar as basic design calculations are concerned. The points where similarities lie are as follows:

Electric circuit	*Magnetic circuit*
Electromotive force, E	Magnetomotive force, F
Current, I	Flux, Φ
Resistance, R	Reluctance, S

The **reluctance** of a magnetic circuit, symbol S, is the effective magnetic 'resistance' to a magnetic flux being established in the magnetic circuit. Electric and magnetic circuits also have a similar 'Ohm's law' to one another as follows:

Electric circuit	*Magnetic circuit*
$E = IR$	$F = \Phi S$

That is, you can consider the m.m.f. in the magnetic circuit to be equivalent to the e.m.f. in the electric circuit, the flux to be equivalent to the current, and the reluctance to be equivalent to the resistance. That is

magnetic circuit 'e.m.f.'	=	magnetic circuit current	×	magnetic circuit 'resistance'

or

m.m.f.	=	flux	×	reluctance

The equation for the reluctance of a piece of magnetic material (see Figure 7.11) is given by

$$\text{reluctance}, S = \frac{1}{\mu a} \text{ ampere-turns per weber}$$

fig 7.11 *Reluctance of a magnetic circuit*

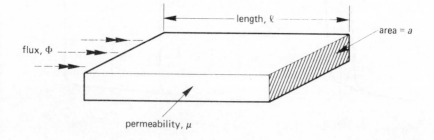

where l is the effective length of the magnetic circuit (in metres), a is the cross-sectional area of the flux path (in m²), and μ is the absolute permeability of the magnetic material.

7.17 MAGNETIC CIRCUITS

A magnetic circuit is simply a path in which magnetic flux can be established. As in electric circuits, you can have series circuits, parallel circuits, series–parallel circuits, etc; the laws which apply to electrical resistance also apply generally to magnetic circuits.

Consider the simple magnetic circuit in Figure 7.12(a) consisting of an iron path with a coil of N turns wound on it. The following values apply to the circuit

number of turns, $N = 1000$
length of the iron path, $l = 0.25$ m
area of iron path, $a = 0.001$ m²
relative permeability of iron path, $\mu_r = 625$
flux in the iron, $\Phi = 1.5 \times 10^{-3}$ Wb

In the following we calculate the value of the current needed in the coil to produce the specified value of flux. The 'equivalent' electrical circuit diagram is shown in Figure 7.12(b) in which S_1 is the reluctance of the magnetic circuit. The value of this reluctance is calculated as follows:

$$\text{reluctance}, S_1 = \frac{l}{\mu_0 \mu_r a}$$

$$= \frac{0.25}{(4\pi \times 10^{-7} \times 625 \times 0.01)}$$

$$= 0.318 \times 10^6 \text{ At/Wb}$$

The m.m.f. produced by the coil is calculated as follows

$$\text{m.m.f.}, F = \Phi S_1 = (1.5 \times 10^{-3}) \times (0.318 \times 10^6)$$

$$= 477 \text{ ampere-turns}$$

But $F = NI_1$, where I_1 is the current in the coil. Hence

$$I_1 = \frac{F}{N} = \frac{477}{1000}$$

$$= 0.477 \text{ A (Ans.)}$$

fig 7.12 *(a) a simple magnetic circuit and (b) its electrically 'equivalent circuit*

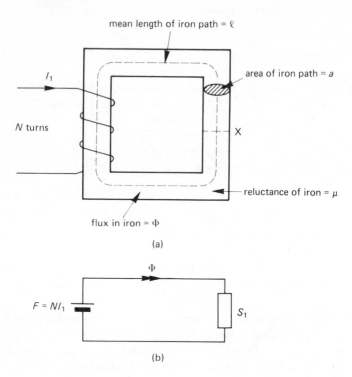

(a)

(b)

A series magnetic circuit

Suppose that the iron path of the magnetic circuit in Figure 7.12(a) is cut at point X so that an air gap, 1 mm-wide, is introduced; this reduces the length of the iron path to 0.249 m. In the following we calculate the current, I_2, which produces the same value of flux (1.5 mWb) in the air gap (see illustrative calculation above). The reluctance of the two parts (the iron path and the air gap) of the magnetic circuit are calculated below (see also Figure 7.13).

Reluctance of iron path, $S_1 = \dfrac{l_{\text{iron}}}{\mu_0 \mu_r a}$

$$= \frac{0.249}{(4\pi \times 10^{-7} \times 625 \times 0.001)}$$

$$= 0.317 \times 10^6 \ \text{At/Wb}$$

fig 7.13 *the 'equivalent' circuit of Figure 7.12(a) but with a 1 mm air gap at X*

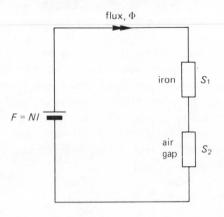

Reluctance of the air gap, $S_2 = \dfrac{l}{\mu_0 a}$ ($\mu_r = 1$ for air)

$$= \frac{0.001}{(4\pi \times 10^{-7} \times 0.001)}$$

$$= 0.796 \times 10^6 \text{ At/Wb}$$

The reader is asked to note the high value of the reluctance of the 1 mm air gap when compared with the reluctance of the 250 mm of iron. The total reluctance of the magnetic circuit is calculated as follows:

Total reluctance, $S = S_1 + S_2 = 1.113 \times 10^6$ At/Wb

The m.m.f. required to produce the flux is

m.m.f., $F = \Phi S = (1.5 \times 10^{-3}) \times (1.113 \times 10^6)$

$$= 1670 \text{ At}$$

The current is calculated from the equation

$$I_2 = \frac{F}{N} = \frac{1670}{1000} = 1.67 \text{ A (Ans.)}$$

The above calculation shows that, in this case, it is necessary to increase the current by a factor of about 3.5 in order to maintain the same magnetic flux in the magnetic circuit when a 1 mm air gap is introduced.

7.18 MAGNETIC SCREENING

Since a strong magnetic field may interfere with the operation of sensitive electrical and electronic apparatus, it may be necessary to screen the apparatus from the field. The method usually adopted is to enclose the apparatus within a screen of low-reluctance material such as iron; this has the effect of placing a magnetic 'short-circuit' around the apparatus.

The screen ensures that the magnetic field cannot penetrate within the shield and reach the apparatus. Magnetic screens are used to protect electrical measuring instruments and cathode-ray tubes from strong fields. In computers, it is possible for a magnetic field produced by, say, a transformer to 'corrupt' the data on a 'floppy disc'; the discs need to be protected against this sort of corruption.

7.19 ELECTROMAGNETIC INDUCTION

When a current flows in a wire, a magnetic flux is established around the wire. If the current changes in value, the magnetic flux also changes in value. The converse is also true, and **if the magnetic flux linking with a wire or coil is changed, then an e.m.f. is induced in the wire or coil**. If the electrical circuit is complete, the induced voltage causes a current to flow in the wire or coil. There are three general methods of inducing an e.m.f. in a circuit, which are

1. self-induction;
2. induction by motion;
3. mutual induction.

The first of the three is fully explained in this chapter, the second and third being only briefly outlined since they are more fully described in Chapters 8 and 14, respectively.

Self induction
If the current in a conductor is, say, increased in value, the magnetic flux produced by the conductor also increases. Since this change of magnetic flux links with the conductor which has produced the fluw, it causes an e.m.f. to be induced in the conductor itself. That is, a change of current in a conductor causes an e.m.f. of **self-induction** to be induced in the conductor.

Induction by motion
An e.m.f. is induced in a conductor when it moves through or 'cuts' a magnetic field. The e.m.f. is due to **induction by motion**, and is the basis of the *electrical generator* (see Chapter 8).

Mutual induction

Suppose that a conductor is situated in the magnetic field of another conductor or coil. If the magnetic flux produced by the other conductor changes, an e.m.f. is induced in the first conductor; in this case the e.m.f. is said to be induced by **mutual induction**. The two conductors or coils are said to be **magnetically coupled** or **magnetically linked**. Mutual inductance is the basic principle of the *transformer* (see Chapter 14).

7.20 THE LAWS OF ELECTROMAGNETIC INDUCTION

The important laws concerned with electromagnetic induction are Faraday's laws and Lenz's law. At this stage in the book we merely state the laws, their interpretation being dealt with in sections appropriate to their practical application.

Faraday's laws

1. **An induced e.m.f. is established in a circuit whenever the magnetic field linking with the circuit changes.**
2. **The magnitude of the induced e.m.f. is proportional to the rate of change of the magnetic flux linking the circuit.**

Lenz's law

> **The induced e.m.f. acts to circulate a current in a direction which opposes the change in flux which causes the e.m.f.**

7.21 SELF-INDUCTANCE OF A CIRCUIT

A circuit is said to have the property of **self-inductance**, symbol L, when a change in the current in its own circuit causes an e.m.f. to be induced in itself. The unit of self-inductance is the **henry**, unit symbol H, which is defined as:

> **A circuit has a self-inductance of one henry if an e.m.f. of one volt is induced in the circuit when the current in that circuit changes at the rate of one ampere per second.**

That is

self-induced e.m.f. = inductance × rate of change of current

or

$$E = \frac{L \times (I_2 - I_1)}{\text{change in time}}$$

or

$$E = L \frac{\Delta I}{\Delta t}$$

where the current in the circuit changes from I_1 to I_2 in the length of time Δt seconds; we use the symbol Δ to represent a change in a value such as time or current.

If the change is very small, that is, it is an *incremental change*, we use the mathematical representation

$$E = L \frac{di}{dt}$$

where $\frac{di}{dt}$ is the mathematical way of waying 'the change in current during the very small time interval dt seconds'.

According to Lenz's law, the self-induced e.m.f. opposes the applied voltage in the circuit; some textbooks account for this by including a negative sign in the equation as follows

$$E = -L \frac{\Delta I}{\Delta t} \text{ and } E = -\frac{Ldi}{dt}$$

However, this can sometimes lead to difficulties in calculations, and in this book the positive mathematical sign is used in the equation for the self-induced e.m.f. . The 'direction' of the induced e.m.f. is accounted for by the way in which we use the induced e.m.f. in the calculation.

The self-inductance of a circuit can simply be thought of as an indication of the ability of the circuit to produce magnetic flux. For the same current in the circuit, a circuit with a low self-inductance produces less flux than one with high inductance.

The self-inductance of a coil with an air core may only be a few millihenrys, whereas a coil with an iron core may have an inductance of several henrys.

Example
Calculate the average value of the e.m.f. induced in a coil of wire of 0.5 H inductance when, in a time interval of 200 ms, the current changes from 2.5 A to 5.5 A.

Solution

$L = 0.5$ H; $\Delta t = 200$ ms $= 0.2$ s;

$\Delta I = 5.5 - 2.5 = 3$ A

Induced e.m.f., $E = L \dfrac{\Delta I}{\Delta t} = 0.5 \times \dfrac{3}{0.2} = 7.5$ V (Ans.)

7.22 RELATIONSHIP BETWEEN THE SELF-INDUCTANCE OF A COIL AND THE NUMBER OF TURNS ON THE COIL

If the number of turns on a coil is increased, then for the same current in the coil, the magnetic flux is increased. That is to say, increasing the number of turns increases the inductance of the coil. The relationship between the self-inductance, L, the number of turns, N, and the reluctance, S, of the magnetic circuit is given by

$$L = \frac{N^2}{S} \text{ H}$$

that is

L is proportional to N^2

If the number of turns of wire on a coil is doubled, the inductance of the coil is increased by a factor of $2^2 = 4$.

7.23 ENERGY STORED IN A MAGNETIC FIELD

During the time that a magnetic field is being established in a magnetic circuit, energy is being stored. The total stored energy is given by

$$\text{energy stored, } W = \frac{1}{2} LI^2 \text{ joules (J)}$$

where L is the inductance of the circuit in henrys and I is the current in amperes.

Example
Calculate the energy stored in the magnetic field of a 3-H inductor which carries a current of 2 A.

Solution

$L = 3\text{H}; \ I = 2\text{A}$

Energy stored, $W = \frac{1}{2} LI^2 = \frac{1}{2} \times 3 \times 2^2 = 6 \text{ J (Ans.)}$

7.24 GROWTH OF CURRENT IN AN INDUCTIVE CIRCUIT

We now study what happens in a circuit when a practical inductor, that is, one having resistance as well as inductance, is connected to an e.m.f., E.

At the instant of time illustrated in Figure 7.14, the blade of switch S is connected to point A; this connection applies a short circuit to the coil, so that the current through the coil is zero. When the blade S of the switch is

fig 7.14 *growth of current in an inductive circuit. The blade of switch S is moving from A to B at the time instant t = 0*

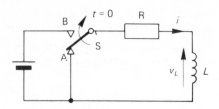

moved to position B (this is described as 'zero' time in this section of the book), the coil (which is an *R–L* circuit) is connected to the battery; at this point in time, current begins to flow through the coil.

When current flows, the inductor produces a magnetic flux. However, this flux 'cuts' its own conductors and induces a 'back' e.m.f. in them (see Faraday's law in section 7.20). The reader will recall from Lenz's law (section 7.20) that this e.m.f. opposes the change which produces it; that is, the e.m.f. opposes the change in the current (which, in this case, is an increase in current). Consequently, the potential arrow v_L associated with the self-induced e.m.f. is shown to oppose the current flow.

Initially, the self-induced 'back' e.m.f. is equal to the supply voltage, E, so that the current in the circuit at time $t = 0$ is

$$i = \frac{\text{supply voltage} - v_L}{R} = \frac{(E - E)}{R} = 0$$

Since this e.m.f. opposes the current, the current does not rise suddenly to a value of $\frac{E}{R}$ (as it does in a purely resistive circuit which contains no inductance) but begins to rise at a steady rate.

The induced e.m.f. is related to the rate of change of current (remember, induced e.m.f. $= L \frac{\Delta I}{\Delta t}$), and as the current increases in value, so its rate of rise diminishes (you can see that the slope of the current graph in Figure 7.15(a) gradually diminishes as the current increases in value). In fact the current builds up according to the relationship

$$i = \frac{E(1 - e^{-t/T})}{R}$$

where $e = 2.71828$ (and is the base of the **natural** or **Naperian logarithms**) and T is the **time constant** of the *R–L* circuit. The time constant of the circuit is calculated from the equation

$$\text{time constant, } T = \frac{L}{R} \text{ seconds}$$

fig 7.15 *(a) rise of current in an inductive circuit and (b) the voltage across the inductor*

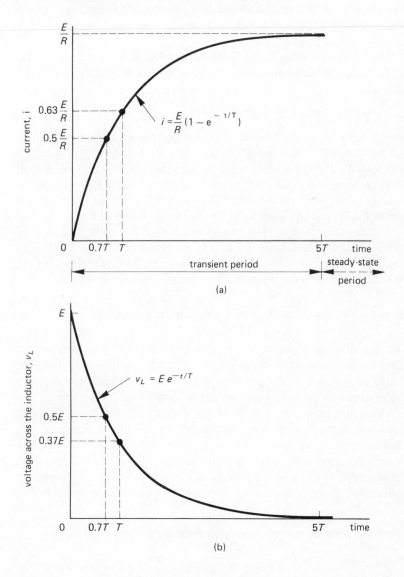

(a)

(b)

where L is the inductance in henrys and R is the resistance in ohms. For example, if $L = 150$ mH or 0.15 H and $R = 10\ \Omega$, then

$$T = \frac{150\,(\text{mH})}{10(\Omega)} = \frac{0.15\,(\text{H})}{10\Omega} = 0.015 \text{ s or } 15 \text{ ms}$$

Figure 7.15(a) shows that the current builds up to 50 per cent of its maximum value in a time of 0.7 of a time constant (which is 10.5 ms for the coil above), and reaches 63 per cent of its final value in a time of one time constant (15 ms in the above case) after the switch blade in Figure 7.14 is changed from A to B.

As the current in the circuit builds up in value, so the voltage across the inductance diminishes according to a decaying exponential curve with the formula

$$\text{voltage across } L, v_L = Ee^{-t/T} \text{ volts}$$

where, once again, e - 2.71828, T is the time constant of the circuit in seconds, and E is the supply voltage. The voltage across the inductive element, L, of the coil diminishes from a value equal to E at the instant that the supply is connected, to practically zero after a time of approximately $5T$ seconds. As shown in Figure 7.15(b), it has fallen to $\frac{E}{2}$ after a time of $0.7T$, and to a value of $0.37E$ after a time equal to T.

As mentioned above, the transient period of the curve is completed after about $5T$ seconds (or $5 \times 15 = 75$ ms in the case where $T = 15$ ms), by which time the current has risen practically to $\frac{E}{R}$ amperes, and the voltage across the inductive element L has fallen practically to zero.

7.25 DECAY OF CURRENT IN AN INDUCTIVE CIRCUIT

We will now study the effect of cutting off the current in an inductive circuit. The circuit is shown in Figure 7.16 and, at a new 'zero' time, the blade of switch S is changed from B to A; this simultaneously applies a short-circuit to the L–R circuit and disconnects it from the battery.

After this instant of time the magnetic flux in the coil begins to decay; however from Faraday' and Lenz's laws we see that a change in flux associated with the circuit induces an e.m.f. in the circuit. According to Lenz's law, the e.m.f. induced in the circuit opposes the change which produces it; in this case, the 'change' is a reduction in current. Clearly, the 'direction' of the induced e.m.f. is *in the same direction as the current* so

fig 7.16 *decay of current in an inductive circuit*

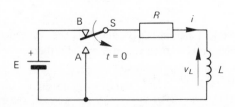

as to try to maintain its value at its original level. Clearly it cannot do this, but has the effect of decaying the fall in the current. Consequently, the 'direction' of the self induced e.m.f. in the coil in Figure 7.16 reverses when compared with its direction in Figure 7.14.

A curve showing the decay in current is illustrated in Figure 7.17(a),

fig 7.17 *waveforms during the current decay in an inductive circuit:*
(a) current in the circuit, (b) voltage across the inductive element

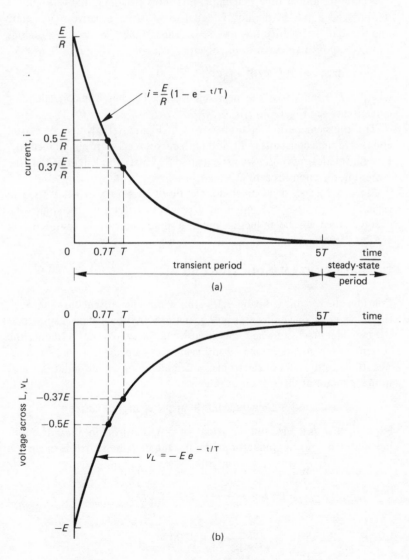

and it follows the exponential law

$$\text{current, } i = \frac{E}{R} e^{-t/T} \text{ A}$$

where $\frac{E}{R}$ is the current in the inductor at the instant that the coil is short-circuited, that is, at $t = 0$; T is the time constant of the circuit $= \frac{L}{R}$ seconds, and $e = 2.71828$. The current decays to half its original value in $0.7T$ seconds and to 37 per cent of its original value in T seconds.

At the instant of time that the coil is short-circuited, that is, $t = 0$, the self-induced e.m.f. in the coil is equal to $-E$ (the negative sign merely implies that its polarity has reversed). This voltage decays to zero along another exponential curve whose equation is

$$\text{voltage across } L, v_L = -E e^{-t/T} \text{ V}$$

where e, E, t and T have the meanings described earlier. The corresponding curve is given in Figure 7.17(b).

The transient period of both curves in Figure 7.17 is complete after about five time constants ($5T$), and both curves reach zero simultaneously. For all practical purposes we may assume that after this length of time the current in the circuit is zero.

Since a special single-pole, double throw (s.p.d.t.) switch is used in Figure 7.16, it does not represent a practical method of discharging the energy stored in the inductive field. A practical case is considered in section 7.26.

7.26 BREAKING AN INDUCTIVE CIRCUIT

Consider the circuit in Figure 7.18(a) in which the current flows through the single-pole, single-throw (s.p.s.t.) switch S. At the instant of time that the switch contact is opened, that is, at $t = 0$, the switch tries to reduce the current in the circuit from a steady value to practically zero in zero time. According to the laws of electromagnetic induction, the self-induced e.m.f. in the inductor at this instant of time is

induced e.m.f. = inductance, L × rate of change of current

Suppose that $L = 3$ H, and a current of 5 A is cut off in 5 ms. The self-induced e.m.f. in the inductor at the instant that the switch is opened is

$$\text{induced e.m.f. } = L \times \text{rate of change of current}$$
$$= L \times \frac{\text{current}}{\text{time}}$$
$$= \frac{3 \times 5}{(5 \times 10^{-3})} = 3000 \text{ V}$$

fig 7.18 *breaking the current in an inductive circuit using a simple switch;*
a spark or arc is produced at the contacts in (b)

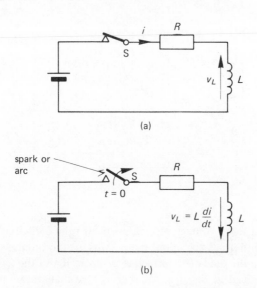

(a)

(b)

At the instant that the switch is opened, the 'direction' of the induced
e.m.f. in the inductor is as shown in Figure 7.18(a) – see also the discussion on Lenz's law in section 7.25. Consequently, **the induced voltage of**
3000 V is added to the supply voltage E, and the combined voltage appears
between the opening contacts of the switch. It is for this reason that a
spark or an arc may be produced at the contacts of the switch when the
current in the inductive circuit is broken. It may be the case that the
supply voltage E has a low value, say 10 or 20 V but, none the less, a
dangerous voltage is developed between the opening contacts of the switch.

In many applications, switches have been replaced by semiconductor
devices such as transistors; these devices are very sinsitive to high voltage,
and are easily damaged. Engineers have therefore looked for ways in which
the high voltage produced by this means can be limited in value.

The common methods used are shown in Figure 7.19. The principle of
all three circuits is the same; that is, when the current through the inductor
is suddenly cut off, an alternative path is provided for the flow of the
inductive current is produced by the decay of the magnetic field.

The inductor is represented in each diagram by R and L in series. In
circuit (a) the inductive circuit is shunted by a capacitor; when the current
in the main circuit is cut off, the energy in the magnetic field is converted
into current i which flows through capacitor C. In circuit (b), the L-R
circuit is shunted by a 'damping' resistor R_D; when the current is cut off
in the main circuit, the induced current produced by the collapse of the

fig 7.19 *methods of absorbing the inductive energy when the current is broken*

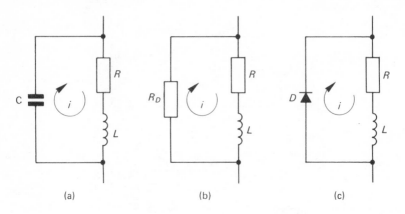

(a)   (b)   (c)

magnetic field flows through R_D. In circuit (c) the L–R circuit is shunted by a diode; in normal operation, the polarity of the supply is such that it reverse-biasses the diode (the principle of operation of the semiconductor diode is described in detail in Chapter 16), so that it does not conduct. When the current through the L–R circuit is cut off, the direction of the induced e.m.f. in the inductor forward-biasses the diode, causing the inductive energy to be dissipated. When used in this type of application, the diode is known as a **flywheel diode** or as a **spark quench diode**.

7.27 **APPLICATIONS OF ELECTROMAGNETIC PRINCIPLES**

A basic application with which everyone is familiar is the **electric bell**. A diagram showing the construction of a bell is **electric bell**. A diagram showing the construction of a bell is illustrated in Figure 7.20, its operation being described below. Initially, when the contacts of the bell push are open, the spring on the iron armature of the bell presses the 'moving' contact to the 'fixed' contact. When the bell push is pressed, the electrical circuit is complete and current flows in the bell coils, energising the electromagnet. The magnetic pull of the electromagnet is sufficiently strong to attract the iron armature against the pull of the spring so that the electrical connection between the fixed and moving contacts is broken, breaking the circuit.

However, the armature is attracted with sufficient force to cause the hammer to strike the gong. Now that the circuit is broken, the pull of the electromagnet stops, and the leaf-spring causes the armature to return to its original position. When it does so, the circuit contact between the fixed and moving contacts is 'made' once more, causing the electromagnet to be

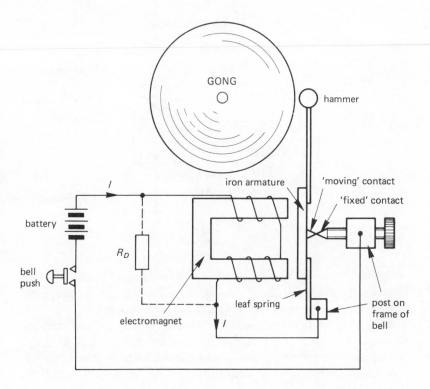

fig 7.20 *an electric bell*

GONG

hammer

iron armature

'moving' contact

'fixed' contact

battery

R_D

bell push

leaf spring

electromagnet

post on frame of bell

energised and the whole process repeated. Only when the bell push is released is the current cut off and the bell stops ringing.

As described earlier, the release of inductive energy when the fixed and moving contacts separate gives rise to a spark between the two contacts. Any one of the methods outlined in Figure 7.19 can be used to limit the sparking and, in Figure 7.20 a damping resistor R_D is shown in dotted line connected across the coil of the bell.

The **relay** is another popular application of electromagnetism (see Figure 7.21). The relay is a piece of equipment which allows a small value of current, I_1, in the coil of the relay to switch on and off a larger value of current, I_2, which flows through the relay contacts.

The *control circuit* of the relay contains the relay coil and the switch S; when S is open, the relay coil is de-energised and the relay contacts are open (that is, the relay has *normally open* contacts). The contacts of the relay are on a strip of conducting material which has a certain amount of 'springiness' in it; the tension in the moving contact produces a downward

fig 7.21 *an electrical relay*

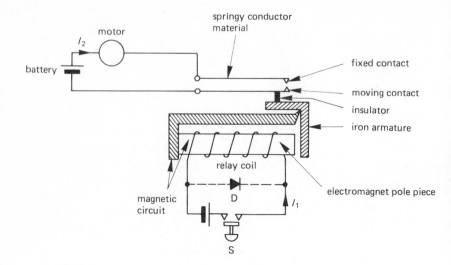

force which, when transferred through the insulating material keeps the iron armature away from the polepiece of the electromagnet.

When switch S is closed, current I_1 flows in the relay coil and energises the relay. The force of the electromagnet overcomes the tension in the moving contact, and forces the moving contact up to the fixed contact. This completes the electrical circuit to the motor, allowing current I_2 to flow in the load.

You might ask why switch S cannot be used to control the motor directly! There are many reasons for using a relay, the following being typical:

1. The current I_1 flowing in the relay coil may be only a few milliamperes, and is insufficient to control the electrical load (in this case a motor which may need a large current to drive it). Incidentally, the switch S may be, in practice, a transistor which can only handle a few milliamperes.
2. The voltage in the control circuit may not be sufficiently large to control the load in the main circuit.
3. There may be a need, from a safety viewpoint, to provide electrical 'isolation' between I_1 and I_2 (this frequently occurs in hospitals and in the mining and petrochemical industries).

Once again, there may be a need to protect the contacts of switch S against damage caused by high induced voltage in the coil when the current

I_1 is broken. The method adopted in Figure 7.21 is to connect a flywheel diode across the relay coil.

Yet another widely-used application of the electromagnetic principle is to the **overcurrent protection** of electrical equipment. You will be aware of the useof the fuse for electrical protection (see Chapter 5), but, in industry, this can be a relatively expensive method of protecting equipment (the reason is that once a fuse is 'blown' it must be thrown away and replaced by a new one). Industrial fuses tend to be much larger and more expensive than domestic fuses.

In industry, fuses are replaced, where possible, by **electromagnetic overcurrent trips**. A simplified diagram of one form is shown in Figure 7.22. The current from the power supply is transmitted to the load via a 'contactor' (which has been manually closed by an operator) and an over-current trip coil. This coil has a non-magnetic rod passing through it which is screwed into an iron slug which just enters the bottom of the overcurrent trip coil; the iron slug is linked to a piston which is an oil-filled cylinder or **dashpot**.

At normal values of load current, the magnetic pull on the iron slug is insufficient to pull the piston away from the drag of the oil, and the contacts of the contactor remain closed.

When an overcurrent occurs (produced by, say, a fault in the load) the current in the circuit rises to a value which causes the magnetic pull produced by the trip-coil to overcome the drag of the oil on the piston. This causes the rod and plunger to shoot suddenly upwards; the top part of the

fig 7.22 *electromagnetic overcurrent trip*

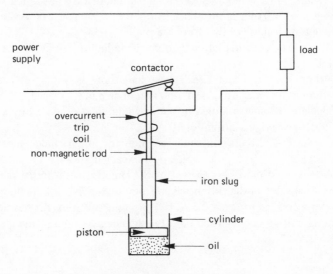

rod hits the contactor and opens the contact to cut off the current to the load. In this way the equipment is protected against overcurrent without the need for a fuse.

The 'value' of the tripping current can be mechanically adjusted by screwing the cylinder and iron slug either up or down to reduce or to increase, respectively, the tripping current.

SELF-TEST QUESTIONS

1. Explain the following terms (i) ferromagnetism, (ii) magnetic domain, (iii) magnetic pole, (iv) magnetic field, (v) direction of the magnetic field, (vi) solenoid, (vii) electromagnet, (viii) magnetic flux.
2. A coil is wound on a non-magnetic ring of mean diameter 15 cm and cross-sectional area 10 cm^2. The coil has 5000 turns of wire and carries a current of 5 A. Calculate the m.m.f. produced by the coil and also the magnetic field intensity in the ring. What is the value of the magnetic flux produced by the coil and the flux density in the ring?
3. Describe the magnetisation curve and the B–H loop of a ferromagnetic material. Explain why a ferromagnetic material 'saturates' when H has a large value. What is meant by 'remanent' flux density and 'coercive force' in connection with the B–H loop?
4. How do magnetically 'soft' and 'hard' materials differ from one another?
5. The relative permeability of a steel ring at a flux density of 1.3 T is 800. Determine the magnetising force required to produce this flux density if the mean length of the ring is 1.0 m.

   What m.m.f. is required to maintain a flux density of 1.3 T if a radial air gap of 1.0 mm is introduced into the ring? You can assume that there is no leakage of flux from the ring.
6. Explain the terms (i) self-induction, (ii) induction by motion and (iii) mutual induction.
7. Outline Faraday's and Lenz's laws and show how they are used to determine the magnitude and direction of an e.m.f. induced in a circuit.
8. Calculate the energy stored in the magnetic field produced by a 2.5 H inductor carrying a current of 10 A.
9. A series R–L circuit containing an inductor of 5 H inductance and a resistance of 10 ohms is suddenly connected to a 10-V d.c. supply. Determine (i) the time constant of the circuit, (ii) the initial value and the final value of the current taken from the supply, (iii) the time taken for the current to reach 0.5 A, (iv) the time taken for the transients in the circuit to have 'settled out', (v) the energy stored in the magnetic field.

SUMMARY OF IMPORTANT FACTS

A **magnetic field** in a **ferromagnetic material** is produced by **magnetic domains**. Lines of magnetic flux are said to **leave a N-pole and enter a S-pole**.

Like magnetic poles attract one another and **unlike magnetic poles repel one another**.

The **magnetomotive force (m.m.f.)** produced by an electromagnet causes a **magnetic flux** to be established in the **magnetic circuit**. The **magnetic field intensity**, H, is the m.m.f. per unit length of the magnetic circuit. The **magnetic flux density**, B, is related to H by the equation

$$B = \mu H$$

where μ is the **permittivity** of the magnetic circuit and

$$\mu = \mu_O \mu_r$$

where μ_O is the permittivity of free space ($= 4\pi \times 10^{-7}$) and u_r is the **relative permittivity** of the material (a dimensionless number).

The **magnetisation curve** for a magnetic material relates B to H for the material. A ferromagnetic material exhibits the property of **magnetic saturation**. The **B-H loop** of the material shows how B and H vary when the magnetising force is increased first in one direction and then in the reverse direction. A **soft** magnetic material has a narrow B-H loop, and a **hard** magnetic material has a flat-fopped 'wide' B-H loop.

The effective 'resistance' of a magnetic circuit to magnetic flux is known as its **reluctance**, S. The relationship between the flux (Φ), the reluctance and the m.m.f. (F) is (**Ohm's law for the magnetic circuit**):

$$F = \Phi S$$

Equipment can be **screened** from a strong magnetic field by surrounding it with a material of low reluctance.

An e.m.f. may be induced in a circuit either by **self-induction**, or **induction by motion in a magnetic field**, or by **mutual induction**. The magnitude and 'direction' of the induced e.m.f. can be predicted using **Faraday's laws** and **Lenz's law**.

The **energy stored**, W, in a magnetic field is given by the equation

$$W = \frac{LI^2}{2}$$

The **time constant**, T, of an L-R circuit is $\frac{L}{R}$ seconds (L in henrys, R in ohms). When an L-R circuit is connected to a d.c. supply, the *final value of the current* in the circuit is $\frac{E}{R}$ amperes, where E is the applied voltage.

The current reaches 63 per cent of its final value after T seconds, and it takes about $5T$ seconds for the transients to 'settle out'. When the voltage applied to the R–L circuit is reduced to zero, it takes T seconds for the current to decay to 37 per cent of its initial value. The transients in the circuit disappear after about $5T$ seconds.

ELECTRICAL GENERATORS AND POWER DISTRIBUTION

8.1 PRINCIPLE OF THE ELECTRICAL GENERATOR

The early activities of scientific pioneers led to the idea that magnetism and electricity were interrelated with one another. It was found that when a permanent magnet was moved towards a coil of wire – see Figure 8.1 – an e.m.f. was induced in the coil. That is, the e.m.f. is induced by the relative movement between the magnetic field and the coil of wire. Lenz's law allows us to predict the polarity of the induced e.m.f.

Lenz's law says that the induced e.m.f. acts to circulate a current in a direction which opposes the change in flux which causes the e.m.f. When the N-pole of the magnet in Figure 8.1 approaches the left-hand end of the coil of wire, the current induced in the coil circulates in a manner to reduce the flux entering the left-hand end of the coil. That is, the current tries to produce a N-pole at the left-hand end of the coil; the effect of the pole produced by the induced current opposes the effect of the approaching magnet. The net result is that the induced current flows in the direction shown in the figure; current therefore *flows out* of terminal Z of the coil (terminal Z is therefore the positive 'pole' of induced e.m.f.).

Consider now the effect of moving the N-pole of the magnet *away* from the coil. When this occurs, the magnetic flux *entering* the left-hand end of the coil is reduced. Lenz's law says that the direction of the induced current is such as to oppose any change. That is, the current induced in the coil must now act to *increase* the amount of flux entering the left-hand end of the coil. The direction of the induced current in the coil must therefore reverse under this condition when compared with the current in Figure 8.1.

If the magnet is repeatedly moved towards the coil and then away from it, the current leaving terminal *Z* flows alternately away from the coil and then towards it (see Figure 8.2). That is to say, **alternating current** (a.c.) is induced in the coil.

fig 8.1   *the principle of electricity generation*

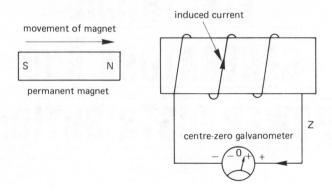

fig 8.2   *a sinusoidal alternating current waveform*

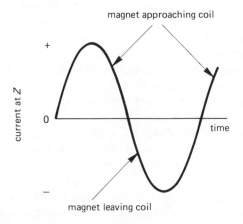

8.2 THE DIRECTION OF INDUCED e.m.f. – FLEMING'S RIGHT-HAND RULE

When studying electrical generators, it is useful to know in which direction the current is urged through a conductor when it moves in a magnetic field. You can predict the direction of the induced e.m.f. by means of **Fleming's right-hand rule** as follows:

> **If the thumb and the first two fingers of the right hand are mutually held at right-angles to one another, and the first finger points in the direction of the magnetic field while the thumb points in the direction of the movement of the conductor relative to the field, then the second finger points in the direction of the induced e.m.f.**

fig 8.3 *Fleming's right-hand rule*

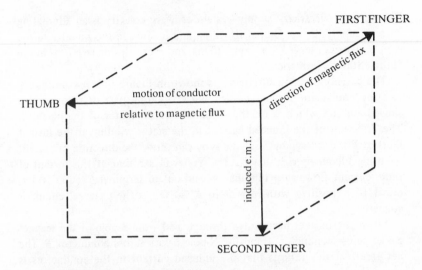

This is illustrated in Figure 8.3 and is summarised below:

*F*irst finger – direction of magnetic *F*lux
s*E*cond finger – direction of induced *E*.m.f.
thu*M*b – *M*otion of the conductor *relative to the flux*.

Examples of the application of Fleming's right-hand rule are given in Figure 8.4.

fig 8.4 *applications of Fleming's right-hand rule*

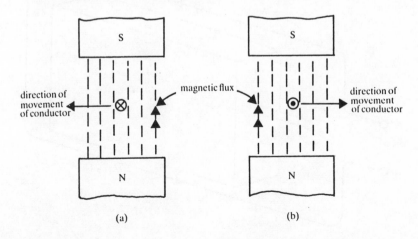

8.3 ALTERNATORS OR a.c. GENERATORS

The national electricity supply system of every country is an alternating current supply; in the United Kingdom and in Europe the polarity of the supply changes every $\frac{1}{50}$ s or every 20 ms, and every $\frac{1}{60}$ s or 16.67 ms in the United States of America.

The basis of a simple alternator is shown in Figure 8.5. It comprises a rotating permanent magnet (which is the rotating part or **rotor**) and a single-loop coil which is on the fixed part or the **stator** of the machine. The direction of the induced current in the stator winding at the instant illustrated is as shown in the figure (you can verify the direction of current by using Fleming's right-hand rule). You will see that at this instant of time, current flows into terminal A' and out of terminal B' (that is, terminal B' is positive with respect to A' so far as the external circuit is concerned).

When the magnet has rotated through 180°, the S-pole of the magnet passes across conductor A and the N-pole passes across conductor B. The net result at this time is that the induced current in the conductors is reversed when compared with Figure 8.5. That is, terminal B' is negative with respect to A'.

In this way, alternating current is induced in each turn of wire on the stator of the alternator. In practice a single turn of wire can neither have enough voltage induced in it nor carry enough current to supply even one

fig 8.5 *a simple single-loop alternator*

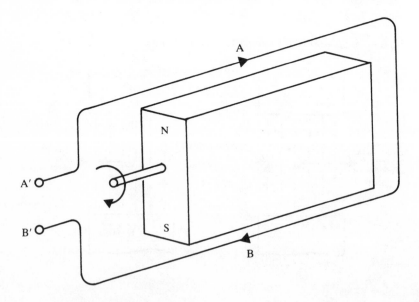

electric light bulb with electricity. A practical alternator has a stator winding with many turns of wire on it, allowing it to deal with high voltage and current. The winding in such a machine is usually *distributed* around the stator in many *slots* in the iron circuit (see Figure 8.6). The designer arranges the coil design so that the alternator generates a voltage which follows a sinewave, that is, the voltage waveform is *sinusoidal* (see Figure 8.7).

The generated e.m.f. rises to a **maximum value** of E_m after 90° rotation. After a further 90° the induced e.m.f. is zero once more and, after a further 90°, the e.m.f. reaches its maximum negative voltage of $-E_m$. A further 90° rotation brings the e.m.f. back to its starting value of zero once more. Figure 8.7 shows one complete **cycle** of the alternating voltage waveform, and the time taken for one complete cycle is known as the **periodic time** of the waveform. The frequency, f, of the wave is the number of complete *cycles per second*, and the frequency of the wave is given in *hertz* (Hz). In the United Kingdom the frequency of the mains power supply is 50 Hz, and the periodic time of the power supply is

$$\text{periodic time} \ = \frac{1}{\text{frequency}} \ (\text{Hz}) = \frac{1}{f}$$

$$= \frac{1}{50} = 0.02 \text{ s or } 20 \text{ ms}$$

Angular frequency

Engineers use the **radian** for angular measurement rather than the degree. One complete cycle (360°) of the wave is equivalent to 2π radians (abbreviated to rad); that is

360° is equivalent to 2π rad

so that

180° is equivalent to π rad

and

90° is equivalent to $\dfrac{\pi}{2}$ rad, etc.

Hence

$$1 \text{ rad} = \frac{360}{2\pi} \text{ degrees} = 57.3°$$

If the frequency of the electricity supply is f Hz or cycles per second, the supply has an **angular frequency**, ω, radians per second of

$$\omega = 2\pi f \text{ rad/s}$$

fig 8.6   *the stator winding of a 3750 kVA alternator*

Reproduced by kind permission of G.E.C. Machines

fig 8.7 *sinusoidal voltage waveform*

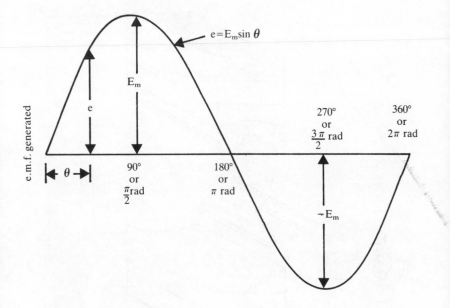

In the UK the angular frequency of the supply is

$$2\pi \times 50 = 314.2 \text{ rad/s}$$

and in the US it is 377 rad/s.

Slip rings

Some alternators use a design which is an 'inversion' of that in Figure 8.5; that is, the magnetic field system is on the stator and the conductors are on the rotor (see Figure 8.8). In this case the field system is an electromagnet excited by a d.c. power supply (in practice it would probably be energised by a d.c. generator known as an **exciter**). The current induced in the rotor winding is connected to a pair of rings, known as **slip rings**, made from either copper, brass or steel; the current is collected from the slip rings by means of carbon brushes which have very little frction and yet have very little voltage drop in them.

In practice, the rotor of the alternator in Figure 8.5 is also an electromagnet which is energised from a d.c. supply via a pair of slip rings.

8.4 SINGLE-PHASE AND POLY-PHASE a.c. SUPPLIES

The alternator described produces a single alternating voltage known as a **single-phase supply**. For a number of reasons which are discussed in

fig 8.8 *a simple alternator with slip rings*

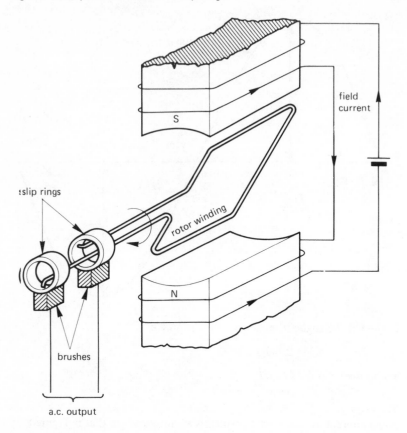

Chapter 13, this type of supply has been replaced for the purpose of national and district power supplies by a **poly-phase supply**.

A poly-phase generator can be thought of as one which has a number of windings on it (usually three windings are involved but strictly speaking, there could be almost any number of windings), each producing its own voltage. This type of generator provides a more economic source of power than several individual single-phase systems.

The National Grid System (both in the UK and other countries) is a three-phase system, and provides the best compromise in terms of overall efficiency.

8.5 EDDY CURRENTS

When a conductor moves through or 'cuts' a magnetic field, an e.m.f. is induced in it. The 'conductor' may, quite simply, be the iron part of a magnetic circuit of a machine such as the rotor of a generator, that is, it may not actually form part of the electrical circuit of the generator. The induced voltage causes a current to flow in the iron 'conductor'; such a current is known as an **eddy current**.

In a number of cases, the 'conductor' may not physically 'move' but may yet have an eddy current induced in it. This occurs in the magnetic circuit of a transformer which carries a pulsating magnetic flux. The pulsation of magnetic flux in the iron induce an eddy current in the iron.

The eddy current simply circulates within the iron of the machine and does not contribute towards the useful output of the machine. Since the resistance of the iron is high, the eddy current causes a power loss (an $I^2 R$ loss) known as the **eddy-current loss**, which is within the general body of the machine and reduces its overall efficiency; in most cases it is clearly to everyone's advantage to reduce eddy currents within the machine. However, there are applications in which eddy currents are used to heat a metal; for example, iron may be melted in an **induction furnace**, in which eddy currents at a frequency of 500 Hz or higher are induced in order to melt the iron.

In the following we study the mechanics of the production of eddy currents; this study leads us to the method used in electrical machines to reduce the induced current.

Consider the solid iron core in Figure 8.9(a) which is wound with a coil or wire carrying an alternating current; at a particular instant of time the current flows in the direction shown and is increasing in value. The direction of the magnetic flux in the iron core is predicted by applying the screw rule (see Chapter 7). Since the excitation current, I, is increasing in value, the magnetic flux in the core is also increasing; the increase in flux in the core induces the eddy current in the iron, giving rise to the eddy-current power-loss.

The eddy current can only be reduced by increasing the resistance of the eddy-current path. The usual way of doing this is to divide the solid iron core into a 'stack' of iron **laminations** as shown in Figure 8.9(b). To prevent eddy-current flow from one lamination to another, they are lightly insulated from one another by either a coat of varnish or an oxide coat. Since the thickness of each lamination is much less than that of the solid core, the electrical resistance of the iron circuit has increased so that the eddy current is reduced in value. There is therefore a significant reduction in the eddy-current power-loss when a laminated magnetic circuit is used.

It is for this reason that the iron circuit of many electrical machines (both d.c. and a.c.) is laminated.

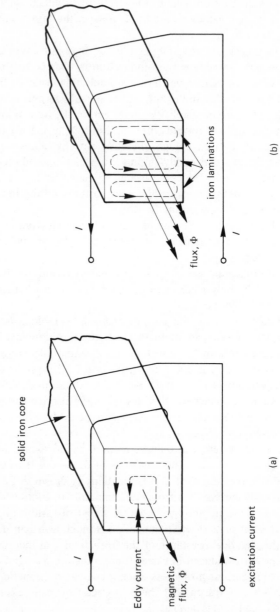

fig 8.9 (a) eddy currents induced in the iron circuit, (b) reduction
of eddy currents by laminating the iron circuit

8.6 DIRECT CURRENT GENERATORS

A **direct current** (d.c.) power supply can be obtained by means of a generator which is generally similar to the alternator in Figure 8.8, the difference between the a.c. and d.c. generators being the way in which the current is collected from the rotating conductors.

Basically, a d.c. generator consists of a set of conductors on the rotating part or **armature** of the d.c. machine, which rotate in the magnetic-field system which is on the fixed part or **frame** of the machine. A diagram of a simple d.c. generator with a single-loop armature is shown in Figure 8.10 (the field winding is excited from a separate d.c. supply [not shown]).

You will see from the diagram that each armature conductor alternately passes a N-pole then a S-pole, so that each conductor has an *alternating voltage* induced in it. However, the current is collected from the conductors by means of a **commutator** consisting of a cylinder which is divided axially to give two segments which enable the alternating current in the conductors to be 'commutated' or 'rectified' into direct current in the external circuit. The way the commutator works is described below.

In the diagram, the conductor WX is connected to the lower segment of the commutator, and the conductor YZ is connected to the upper segment. At the instant of time shown, the e.m.f. in the armature causes current to flow from W to X and from Y to Z; that is, current flows out of the upper commutator segment and into the lower commutator segment.

fig 8.10 *a simple single-loop d.c. generator*

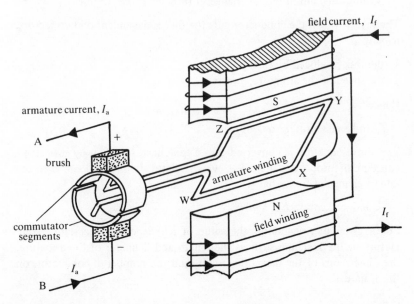

When the armature has rotated through 180°, the conductor WX passes the S-pole of the field system and conductor YZ passes the N-pole. This causes the current in the armature conductors to reverse but, at this time, the position of the commutator segments have also reversed so that the current once again flows out of the upper segment and into the lower segment.

The function of the commutator in a d.c. machine is to ensure that the external circuit is connected via the brushgear to a fixed point in space inside the armature, i.e., to a point of fixed polarity.

In Figure 8.10, point A in the external circuit is *always* connected to a conductor in the armature which is moving under the S-pole, and point B is *always* connected to a conductor which is moving under the N-pole.

In practice the armature winding is a coil having many hundreds of turns of wire which is tapped at several dozen points, each tapping point being connected to an individual commutator segment. Each segment is separated from the next one by an insulating material such as mica.

8.7 SIMPLIFIED e.m.f. EQUATION OF THE d.c. GENERATOR

As mentioned earlier, the e.m.f. induced in a conductor in a generator is proportional to the rate at which it cuts the magnetic flux of the machine. That is:

induced e.m.f. α rate of change of flux

The rate at which the conductor cuts the flux is dependent on two factors, namely:

1. the flux produced by each pole, Φ
2. the speed of the armature, n rev/s

Hence

induced e.m.f., $E \ \alpha \ \Phi n$

The above equation is converted into a 'practical' equation by inserting a constant of proportionality, K_E, which relates the induced e.m.f. to the rate of change of flux. That is

induced e.m.f., $E = K_E \Phi n$

For a given d.c. generator, the value of K_E depends on certain 'fixed' factors in the machine such as the length and diameter of the armature, the dimensions of the magnetic circuit, and the number of conductors on the armature.

8.8 AN ELECTRICITY GENERATING STATION

The basis of an electrical generating plant is shown in Figure 8.11. The power station is supplied with vital items such as water and fuel (coal, oil, nuclear) to produce the steam which drives the turbine round (you should note that other types of turbine such as water power and gas are also used). In turn, the turbine drives the rotor of the alternator round. As shown in the figure, the rotor of the alternator carries the field windings which are excited from a d.c. generator (which is mechanically on the same shaft as the alternator) via a set of slip rings and brushes.

The stator of the alternator has a three-phase winding on it, and provides power to the transmission system. The voltage generated by the alternator could, typically, be 6600 V, or 11000 V, or 33000 V.

fig **8.11** *a simplified diagram of en electrical power station*

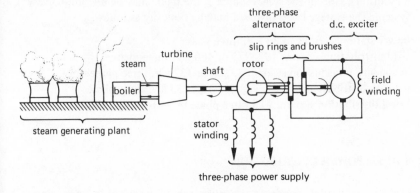

8.9 THE a.c. ELECTRICAL POWER DISTRIBUTION SYSTEM

One advantage of an a.c. supply when compared with a d.c. supply is the ease with which the voltage level at any point in the system can be 'transformed' to another voltage level. The principle of operation of the electrical transformer is described in detail in Chapter 14 but, for the moment, the reader is asked to accept that it is a relatively easy task to convert, say, a 6.6 kV supply to 132 kV, and vice versa.

In its simples terms, electrical power is the product of voltage and current and, if the power can be transmitted at a high voltage, the current is correspondingly small. For example, if, in system A, power is transmitted at 11 kV and, in system *B*, it is transmitted at 33 kV then, for the same amount of power transmitted, the current in system A is three times greater than that in system B. However, the story does not finish there because:

1. the voltage drop in the transmission lines is proportional to the current in the lines;
2. the power loss in the resistance of the transmission lines is proportional to (current)^2 [remember, power loss = I^2R].

Since the current in system A is three times greater than that in system B, the voltage drop in the transmission lines in system A is three times greater than that in system B, and the power loss is nine times greater!

This example illustrates the need to transmit electrical power at the highest voltage possible. Also, since alternating voltages can easily be transformed from one level to another, the reason for using an a.c. power system for both national and local power distribution is self-evident.

The basis of the British power distribution system is shown in Figure 8.12 (that of other countries is generally similar, but the voltage levels may differ). Not all the stages shown in the figure may be present in every system. The voltage levels involved may be typically as follows:

Power station: 6.6, 11 or 33 kV, three phase
Grid distribution system: 132, 275 or 400 kV, three phase
Secondary transmission system: 33, 66 or 132 kV, three phase
Primary distribution system: 3.3, 6.6, 11 or 33 kV, three phase
Local distribution system: 415 three-phase or 240 V single phase.

8.10 d.c. POWER DISTRIBUTION

For certain limited applications, power can be transmitted using direct current. The advantages and disadvantages of this when compared with a.c. transmission are listed below.

Advantages
1. A given thickness of insulation on cables can withstand a higher direct voltage than it can withstand alternating voltage, giving a smaller overall cable size for d.c. transmission.
2. A transmission line has a given cable capacitance and, in the case of an a.c. transmission system this is charged continuously. In the case of a d.c. transmission system, the charging current only flows when the line is first energised.
3. The self-inductance of the transmission line causes a voltage drop when a.c. is transmitted; this does not occur when d.c. is transmitted.

fig 8.12 *the national power distribution system*

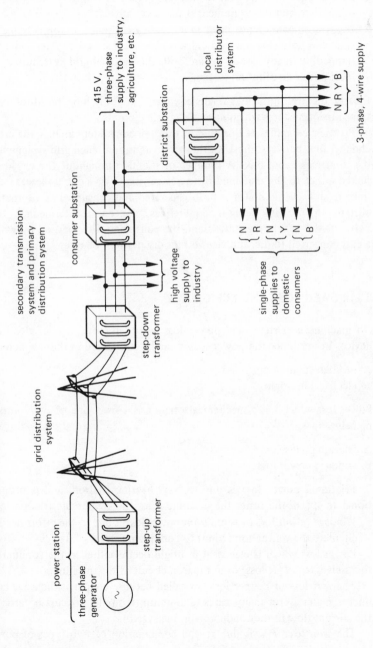

Disadvantages

1. Special equipment is needed to change the d.c. voltage from one level to another, and the equipment is very expensive.
2. D.c. transmission lends itself more readily to 'point-to-point' transmission, and problems arise if d.c. transmission is used on a system which is 'tapped' at many points (as are both the national grid system and the local power distribution system.

Clearly, d.c. transmission is financially viable on fairly long 'point-to-point' transmission systems which have no 'tapping' points.

Practical examples of this kind of transmission system include the cross-channel link between the UK grid system and the French grid system via a d.c. undersea cable link. A number of islands throughout the world are linked either to the mainland or to a larger island via a d.c. undeasea cable link. In any event, power is both generated and consumed as alternating current, the d.c. link being used merely as a convenient intermediate stage between the generating station and the consumer. The method by which a.c. is converted to d.c. and vice versa is discussed in Chapter 16.

8.11 POWER LOSS AND EFFICIENCY

No machine is perfect, and power loss occurs in every type of electrical device. In machines the power losses can be considered as those which are

1. mechanical in origin
2. electrical in origin.

Power loss which has a **mechanical origin** arises from one of two sources, namely

A. friction power loss
B. windage power loss.

Frictional power loss is due to, say, bearing friction, and is proportional to speed; the faster the machine runs, the greater the friction loss.

Windage power loss or *ventilation* power loss is due to the effort needed to circulate the ventilation (wind) to cool the machine.

Power loss which is **electrical in origin** occurs either in the conductors themselves (the I^2R loss) or in the iron circuit of the machine.

The I^2R loss or **copper loss** (so called because copper is the usual conductor material [in many cases, aluminium is used!]) occurs because of the current flow in the conductors in the system.

The **iron loss**, P_O, is due to the combination of two types of power loss:

1. the **hysteresis power-loss**, P_n, occurs because the state of the magnetisation of certain parts of the magnetic circuit is reversed at regular intervals; work is done in reversing the magnetic 'domains' when this occurs;
2. the **eddy-current loss**, which is due to the heat generated by the flow of eddy currents in the iron circuit.

The iron loss is also known as the **core loss**, P_C. If the supply voltage and frequency are constant, the iron loss is a *constant power loss* and is independent of the load current.

8.12 CALCULATION OF THE EFFICIENCY OF A MACHINE

The efficiency, symbol η, of an electrical machine is the ratio of the electrical power output from the machine to its power input, as follows

$$\text{efficiency, } \eta = \frac{\text{output power}}{\text{input power}}$$

However,

output power = input power − power loss

and

input power = output power + power loss

where

power loss = electrical power loss + mechanical power loss

$$= I^2R \text{ loss} + P_O \text{ loss} + \text{friction loss} + \text{windage loss}$$

Note: not all of the above losses appear in every electrical machine (for instance, the transformer does not have any moving parts and does not, therefore, have either friction or windage losses). The efficiency of an electrical machine can therefore be expressed in one of the following forms:

$$\text{efficiency, } \eta = \frac{\text{output power}}{\text{output power} + \text{losses}}$$

$$= \frac{\text{input power} - \text{losses}}{\text{input power}}$$

In both the above equations, the denominator is greater than the numerator, so that the efficiency of electrical machines can never be 100 per cent (although in many cases, the efficiency is very high).

The result of each of the above equations is a value which is less than unity; for this reason the basic unit of efficiency is expressed as a **per unit** or p.u. value. For example, if the output power from a generator is 90 kW and the mechanical input power is 100 kW, the efficiency is

$$\text{efficiency}, \eta = \frac{\text{output power}}{\text{input power}}$$

$$= \frac{90}{100} = 0.9 \text{ per unit or p.u.}$$

It is sometimes convenient to express the efficiency in a **per cent** form as follows

per cent efficiency = per unit efficiency × 100

so that a 0.9 per unit efficiency corresponds to a 90 pr cent efficiency.

Example
An electrical generator provides an output power of 270 kW. The electrical losses in the machine are found to be 36 kW and the friction and windage loss is 12 kW. Calculate the mechanical input power to the machine and its overall efficiency.

Solution

Mechanical input power	=	electrical output power	+	electrical power loss	+	mechanical power loss

$$= 270 + 26 + 12 = 318 \text{ kW (Ans.)}$$

and

$$\text{Efficiency}, \eta = \frac{\text{output power}}{\text{input power}}$$

$$= \frac{270}{318}$$

$$= 0.849 \text{ p.u. or } 84.9 \text{ per cent (Ans.)}$$

SELF-TEST QUESTIONS

1. Draw a diagram of an alternator and explain its operation. If the alternator has 4 poles and its speed of rotation is 6000 rev/min, calculate the frequency of the a.c. supply.

2. Convert (i) 60° and (ii) 120° into radians and convert (iii) 1.6 radians and (iv) 4.6 radians into degrees.
3. Compare the relative merits of single-phase and three-phase supplies.
4. What is meant by an 'eddy current'? How are eddy currents reduced in value in an electrical machine?
5. Explain the principles of operation of a d.c. generator. Compare the functions of 'slip rings' and a 'commutator'.
6. Draw a simplified diagram of an electricity generating station and explain its operation. Draw also a diagram of the power distribution network and describe the purpose of each section of the distribution system.
7. Compare a.c. and d.c. distribution systems.
8. Write down a list of power losses which occur in an electrical machine and explain how it affects the efficiency of the machine.

SUMMARY OF IMPORTANT FACTS

An **alternator** or **a.c. generator** has a rotating part known as a **rotor** and a stationary part or **stator**. The corresponding parts of a d.c. generator are the **armature** and **frame**, respectively. The direction of the current in the conductors can be predicted by **Fleming's right-hand rule**. Current is collected from the **commutator** of a d.c. generator.

The **frequency**, f, in hertz (Hz) of an alternating wave is related to the **periodic time**, T, in seconds by the equation $f = \frac{1}{T}$.

Angles are measured in either **degrees** or **radians**, the two being related as follows:

$$\text{radians} = 2\pi \times \frac{\text{degrees}}{360}$$

Angular frequency, ω in rad/s is given by

$$\omega = 2\pi f \text{ rad/s}$$

where f is the supply frequency in Hz.

Power is distributed through the **National Grid** system using a **three-phase, three-wire** system, and is distributed to domestic consumers using a **three-phase, four-wire** system; each domestic consumer uses power taken from **one phase** of this system. Power is transmitted across short sea passages such as the English Channel using **high voltage direct current** distribution.

An **eddy current** is a current which is induced in the **iron circuit** of an electrical machine (which could be either an a.c. machine or a d.c. machine) whenever the magnetic flux linking with the iron changes. This gives rise to an **eddy-current power-loss** in the iron, causing it to heat up. Eddy currents are minimised by using a **laminated** iron circuit.

Power loss in an electrical machine may either be **mechanical** in origin, that is, due to **friction** or **windage**, or it may be **electrical** in origin, that is, due either to I^2R **loss**, or to **hysteresis loss**, or to **eddy-current loss**. The **efficiency** of an electrical machine is given by the ratio of the output power to the input power and is either expressed in **per unit** or in **per cent**.

DIRECT CURRENT
MOTORS

9.1 THE MOTOR EFFECT

The **motor effect** can be regarded as the opposite of the **generator effect** as follows. In a generator, when a conductor is moved through a magnetic field, a current is induced in the conductor (more correctly, an e.m.f. is induced in the conductor, but the outcome is usually a current in the conductor). In a motor, a current-carrying conductor which is situated in a magnetic field experiences a force which results in the conductor moving (strictly speaking, the force is on the *current and not on the conductor*, but the current and the conductor are inseparable).

9.2 THE DIRECTION OF THE FORCE ON THE CURRENT-CARRYING CONDUCTOR – FLEMING'S LEFT-HAND RULE

Fleming's left-hand rule allows you to predict the direction of the force acting on a current-carrying conductor, that is, it allows you to study the effect of *motor* action (Fleming's left-hand rule is for **motor action** [you can remember which rule is for motors by the fact that in the UK *motors drive on the left hand side of the road*]). The left-hand rule is as follows:

If the thumb and first two fingers of the left hand are mutually held at right-angles to one another, and the first finger points in the direction of the magnetic field whilst the second finger points in the direction of the current, then the thumb indicates the direction of the force on the conductor.

This is illustrated in Figure 9.1 and can be summarised by:

*F*irst finger – direction of the magnetic *F*lux
se*C*ond finger – direction of the *C*urrent
thu*M*b – direction of the *M*otion of (or force on) the conductor.

fig 9.1 *Fleming's left-hand rule*

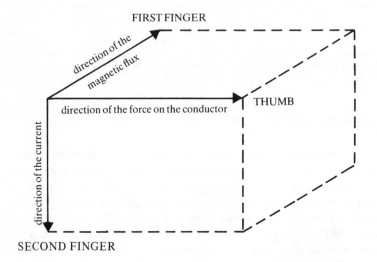

An application of Fleming's left-hand rule is given in Figure 9.2.

fig 9.2 *an application of Fleming's left-hand rule*

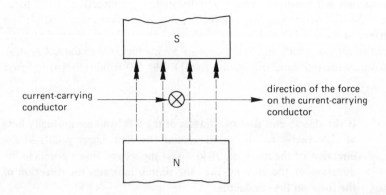

9.3 THE FORCE ON A CURRENT-CARRYING LOOP OF WIRE

Suppose that the loop of wire WX in Figure 9.3 is pivoted about its centre and carries a direct current in the direction shown. Fleming's left-hand rule shows that the force on wire W is in the opposte direction to that on wire X, the combined effect of the two forces causing the loop to rotate in a clockwise direction.

fig 9.3 *force on a current-carrying loop of wire*

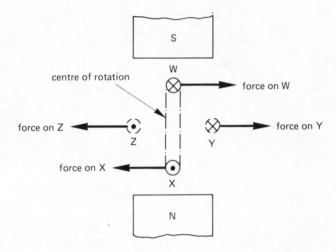

The loop of wire continues to rotate about its centre until it reaches position YZ. At this point, the forces on the two conductors are not only opposite in direction but are in the same plane of action. The net **torque** or mechanical **couple** acting on the coil is zero, and it stops rotating. Clearly, to obtain continuous rotation, the circuit arrangement must be modified in some way; the method adopted is described in section 9.4.

9.4 THE d.c. MOTOR PRINCIPLE

A simple single-loop d.c. motor is shown in Figure 9.4. In the motor shown, the magnetic **field system** is fixed to the frame of the motor, and the rotating part or **armature** supports the current-carrying conductors. The current in the field coils is known as the **excitation current** or **field current**, and the flux which the field system produces reacts with the **armature current** to produce the useful mechanical output power from the motor.

fig 9.4 *the basis of a d.c. motor*

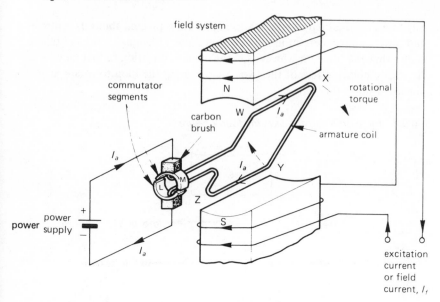

The armature current is conveyed to the armature via carbon brushes and the commutator (the function of the armature having been described in Chapter 8). It is worthwhile at this point to remind ourselves of the functions of the **commutator**. First, it provides an electrical connection between the armature winding and the external circuit and, second, it permits reversal of the armature current whilst allowing the armature to continue to produce a torque in one direction.

At the instant shown in Figure 9.4, current flows from the positive terminal of the power supply into the armature conductor WX, the current returning to the negative pole of the supply via conductor YZ. The force on the armature conductors causes the armature to rotate in the direction shown (the reader might like to verify this using Fleming's rule).

When the armature winding reaches the horizontal position, the gap in the commutator segments passes under the brushes so that the current in the armature begins to reverse. When the armature has rotated a little further, conductor WX passes under the S-pole and YZ passes under the N-pole. However, the current in these conductors has reversed, so that the torque acting on the armature is maintained in the direction shown in Figure 9.4. In this way it is possible to maintain continuous rotation.

9.5 CONSTRUCTION OF A d.c. MOTOR

A cross section through a d.c. motor is shown in Figure 9.5. The **frame** of the machine consists of an iron or a steel ring known as a **yoke** to which is attached the main **pole system** consisting of one or more *pairs of poles*. Each pole has a **pole core** which carries a **field winding**, the winding being secured in position by means of a large soft-iron **pole tip**. Some machines have a smaller pole known as an **interpole** or a **compole** between each pair of main poles; the function of the compoles is to improve the armature current commutation, that is, to reduce the sparking at the brushes under conditions of heavy load.

To reduce the eddy-current power-loss in the armature, the cylindrical **armature** is constructed from a large number of thin iron *laminations*. The armature conductors are carried in, but insulated from, teeth cut in the armature laminations.

fig 9.5 *d.c. machine construction*

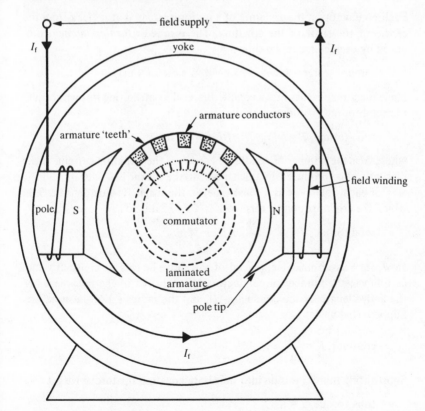

9.6 MAGNITUDE OF THE FORCE ON A CURRENT-CARRYING CONDUCTOR

It has been found experimentally that the force, F, acting on a current-carrying conductor placed in a magnetic field is given by

$$\text{force, } F = BIl \text{ newtons (N)}$$

where B is the flux density (tesla) produced by the field system, I is the current (amperes) in the conductor, and l is the 'active' length in metres (m) of the conductor in the magnetic field.

For example, if a conductor of length 0.3 m carries a current of 100 A in a magnetic field of flux density 0.4 T, the force on the conductor is

$$\text{force, } F = BIl = 0.4 \times 100 \times 0.3 = 12 \text{ N}$$

(strictly speaking the force is on the current).

9.7 TORQUE PRODUCED BY THE ARMATURE OF A d.c. MOTOR

Each conductor on the armature of a motor is situated at radius r from the centre of the shaft of the armature. The **torque** or *turning moment* produced by each conductor is shown by

$$\text{torque} = \text{force} \times \text{radius} = Fr \text{ newton metres (N m)}$$

Since the armature has many conductors, each contributing its own torque, the total torque T produced by the armature is

$$\text{total torque, } T = NFr = N \times BI_a \, l \times r \text{ N m}$$

where N is the number of conductors and I_a is the armature current. If the magnetic flux produced by each pole of the motor is Φ webers and the area of each pole-piece is a square metre, the magnetic flux density, B, in which the armature conductors work is $\frac{\Phi}{a}$ T. That is to say

$$\text{total torque, } T = N \times \ \frac{\Phi}{a} \ I_a l \times r \text{ N m}$$

Now, for a given machine, several of the above factors are constant values as follows; the number of conductors (N), the area of the pole-piece (a), the active length of the conductor (l), and the radius (r) of the armature. Suppose that we let

$$\text{constant, } K = \frac{Nlr}{a}$$

Substituting this expression into the above equation for torque we get

$$\text{total torque, } T = K\Phi I_a$$

That is

total torque is proportional to ΦI_a

or

$T \alpha \Phi I_a$

From this it can be seen that the primary factors affecting the torque produced by the motor are the flux per pole (Φ) and the armature current (I_a). Increasing either of these results in an increased torque.

9.8 'BACK' e.m.f. INDUCED IN A d.c. MOTOR ARMATURE

When the armature of a d.c. motor rotates, the conductors cut the magnetic field of the machine. From the earlier work in this book, you will appreciate that when a conductor cuts a magnetic flux, an e.m.f. is induced in the conductor. That is, even though the machine is acting as a motor, there is an e.m.f. induced in the armature conductors. However, in the case of a d.c. motor, the induced e.m.f. opposes the flow of current through the motor (and is another illustration of Lenz's law). For this reason, the e.m.f. induced in the armature of a d.c. motor is known as a **back e.m.f.**

This can be understood by reference to Figure 9.6. The generator armature is driven by a steam turbine, and the generated voltage is E_1; suppose that this voltage is 500 V. When the motor is driving its load, its armature has a 'back' e.m.f., E_2, induced in it. For a current to circulate from the generator to the motor, the value of E_1 must be greater than E_2. Suppose that E_2 is 490 V; the potential difference, $(E_1 - E_2) = 10$ V, is used up in overcoming the resistance of the circuit in order to drive current through it.

fig 9.6 *motor 'back' e.m.f.*

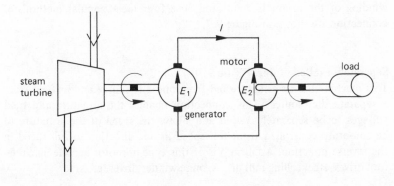

If the value of the armature current I is 10 A, the power generated by the generator armature is

$E_1 I = 500 \times 10 = 5000$ W

The power consumed by the motor armature is

$E_2 I = 490 \times 10 = 4900$ W

The difference between the two power values of

$(5000 - 4900) = 100$ W

is consumed in the resistance of the circuit.

In practice, when a d.c. motor is *running at full speed*, the armature back e.m.f. is slightly less than that of the supply voltage. You should note that this applies at full speed only; if the speed is less than full speed, then the back e.m.f. is also less (this is discussed further in section 9.11).

Since the armature conductors have a 'back' e.m.f. induced in them, you can think of this e.m.f. as being a 'generated' e.m.f. (albeit a 'back' e.m.f.). From this point of view, the relationship between the back e.m.f. and other factors in the machine are the same for the motor as for the generator. We can therefore use the 'generator' equation for the back e.m.f. of the motor as follows:

back e.m.f. $= K_E \Phi n$

where K_E is a constant of the machine, Φ is the magnetic flux per pole of the motor, and n is the armature speed. Later we will use this equation to show how speed control of the motor can be achieved.

9.9 TYPES OF d.c. MOTOR

The 'name' or 'type' of d.c. motor is given to the way in which the field winding of the motor is connected. The four most popular methods of connection are shown in Figure 9.7.

Separately excited motor (Figure 9.7(a))
This machine has its armature fed from one d.c. source and its field from a separate d.c. source. This connection enables the armature and field voltages to be separately varied and allows the speed of the armature to be smoothly controlled from full speed in one direction to full speed in the reverse direction. Applications of this type of motor include machine-tool drives, steel rolling-mill drives, mine-winder drives, etc.

fig 9.7 *d.c. motor connections*

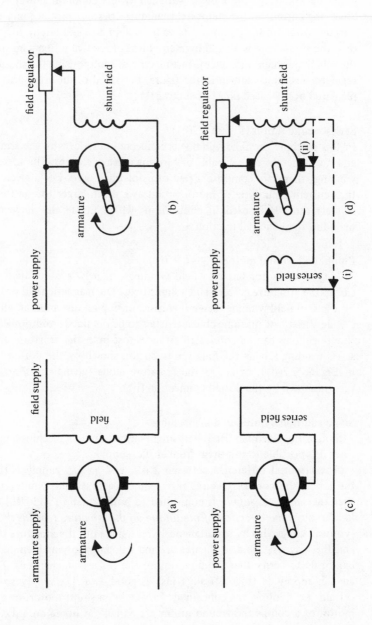

Shunt-wound motor (Figure 9.7(b))

In this type of motor, the field winding is connected in shunt or in parallel with the armature, both being supplied from a common power supply. Since the current in the shunt-field winding does not produce any useful output power from the motor, its value is kept to a minimum; to achieve this, the shunt field is wound by many turns of relatively fine wire (to give the winding a high resistance). In this type of motor; using the shunt-field **regulator** resistance shown in the figure, the speed of the armature can be regulated over a speed range of about 3:1.

Series-wound motor (Figure 9.7(c))

In this machine the field winding is connected in series with the armature, and must therefore be capable of carrying a heavy current. The series-field windings therefore comprise a few turns of thick wire. For a given armature current, this type of motor produces a much larger torque than the shunt motor. It is used in applications which utilise this large torque including road and rail traction.

Compound-wound motor (Figure 9.7(d))

This type of motor has two field windings, namely a series field which carries the armature current and a shunt field. The magnetic field produced by the two field windings can either assist or oppose one another, allowing a wide variety of machine characteristics to be obtained. Additionally, the shunt winding can be connected either across both the armature and the series winding (giving the *long shunt* version shown by the dotted line (i) in Figure 9.7(d)), or across the armature alone (giving the *short shunt* version shown by the dotted connection (ii)).

Some additional notes on d.c. machines

Although the machines listed here are described as 'd.c.' machines, in some cases it is possible to run them from an a.c. supply.

The essential difference between a d.c. and an a.c. supply is that, in the latter, the current reverses many times each second. Now, provided that the *same a.c. supply* is connected to both armature and field of the 'd.c.' motor, *the direction of the torque on the armature remains the same* (you can check this by simultaneously reversing both the armature current and the magnetic field in Figures 9.2 and 9.3, and applying Fleming's left-hand rule to verify that the direction of the force is unchanged). That is, an a.c. supply is (theoretically) just as good as a d.c. supply so far as certain d.c. motors are concerned. In general, a shunt motor, or a series motor, or a compound-wound motor are equally at home on a d.c. or an a.c. supply.

A problem associated with **large series motors** which does not occur in other d.c. motors is that, if the mechanical load is disconnected from its shaft, the armature 'races'and can reach a dangerously high speed. It is for this reason that large series motors often have an additional shunt-winding which limits the armature speed to a safe maximum value.

Many handtools such as drills and woodworking tools are driven by series motors. In general, the friction and windage power-loss in these machines is a very high proportion of the total load, so that the problem of the armature 'racing' when the load is disconnected is non-existent.

9.10 COMMUTATION PROBLEMS IN d.c. MACHINES

The current which causes the armature to produce its mechanical output power has to flow through the brushes and commutator before reaching the armature. Since the contact between the two is simply a mechanical one, there is a risk of electrical sparking at the point of contact. Several methods are adopted to reduce the sparking to a minimum and include:

1. the use of brushes whose resistance is not zero (carbon brushes are used since they have about the correct resistance);
2. the use of brushes which span several commutator segments;
3. the use of *interpoles* or *commutating poles* (abbreviated to *compoles*) between the main poles. These poles induce a voltage in the armature which combats the sparking effect (compoles are fairly expensive, and are used only in large machines).

9.11 MOTOR STARTERS

A d.c. motor with a rating above about 100 watts has a very low armature resistance (electrical handtools have a fairly low power-rating and have a fairly high armature-circuit resistance). Consequently, when the d.c. supply is switched on, the current drawn by the armature under starting conditions can be very high. Unless the starting current can be limited to a safe value, there is a considerable dange of the motor being damaged by

1. excessive heating (due to I^2R) in the armature;
2. sparking between the brushes and the commutator at the point of contact.

The simplest method of reducing the starting current is to insert resistor R (see Figure 9.8(a)) in series with the armature. Under starting conditions, the armature is stationary and the 'back' e.m.f. in the armature is zero. The p.d. across resistor R under starting conditions is therefore practically the whole of the supply voltage. If, for example, the supply voltage is

fig 9.8 *simplified starter for a d.c. shunt motor*

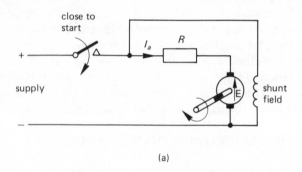

(a)

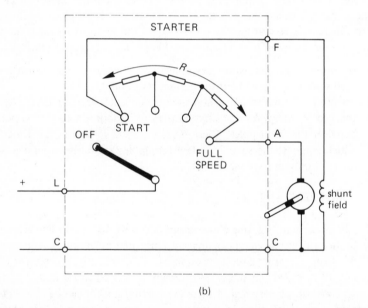

(b)

240 V and the starting current must be limited to 10 A, the value of R is calculated from

$$R = \frac{V_S}{I} = \frac{240}{10} = 24 \ \Omega$$

This current causes the armature to produce a torque, so that it accelerates from standstill. As it does, a back e.m.f. E is induced in the armature. Since this e.m.f. opposes the supply voltage, the p.d. across R (= $V_S - E$)

decreases. For example, when E = 100 V, the p.d. across R is $(240 - 100)$ = 140 V; the armature current is then reduced to

$$\frac{(V_S - E)}{R} = \frac{140}{24} = 5.83 \text{ A}$$

Since the armature can safely withstand a surge of current of 10 A, it is possible to reduce the value of R to allow more current to flow, and to enable the armature to accelerate further.

One way in which this is done in a manually-operated starter is shown in Figure 9.8(b). The starting resistance R is divided into several steps; the arm of the starter is moved from the OFF position to the START position, and is then slowly moved from one stud to the next as the starting resistance is cut out of the armature circuit. When full speed is reached, the starting resistance is completely cut out of the circuit, and the supply voltage is connected directly to the armature. At this point in time, the back e.m.f. E in the armature has a value which is only slightly less than that of the supply voltage.

Although not shown in the starter circuit in Figure 9.8(b), a practical starter would incorporate circuits which protect the motor not only against *overcurrent* but also against *undervoltage* during normal operation.

Automatic starters are available which automatically reduce the starting resistance either on a timing basis or on a current value basis.

SELF-TEST QUESTIONS

1. Explain how Fleming's left-hand rule can be used to determine the direction of the current induced in a conductor.
2. Explain what is meant by the following terms in association with a d.c. machine: armature, commutator, brushes, field system, frame, interpole.
3. An armature conductor carries a current of 86 A. If the length of the conductor is 0.25 m and the force on the conductor is 8 N, calculate the flux density in the machine.
4. The torque produced by a motor is 1000 N m. If the magnetic flux is reduced by 20 per cent and the armature current is increased by 20 per cent, what torque is produced?
5. Name four types of d.c. machine and discuss applications for each of them.
6. Discuss the need for a d.c. motor starter. Draw a circuit diagram of a motor starter and explain its operation.

SUMMARY OF IMPORTANT FACTS

Motor action is caused by the force acting on a current-carrying conductor in a magnetic field. The direction of the force can be predicted by **Fleming's left-hand rule**.

A **d.c. motor** consists of a *rotating part* (the **armature**) and a *fixed part* (the **frame**). Electrical connection to the armature is made via **carbon brushes** and the **commutator**. The **torque** produced by the armature is proportional to the *product of the field flux and the armature current*. When the armature rotates, a **back e.m.f.** is induced in the armature conductors (this is by *generator action*) which opposes the applied voltage.

The four main types of d.c. motor are the **separately excited**, the **shunt wound**, the **series wound** and **compound wound** machines.

d.c. machines experience **commutation problems**; that is, sparking occurs between the brushes and the commutator. These problems can be overcome, in the main, by using brushes which have a finite resistance and which span several commutator segments (wide *carbon brushes*) together with the use of *interpoles* or *compoles*.

d.c. motors larger than about 100 W rating need a **starter** in order to limit the current drawn by the motor under starting conditions to a safe value.

ALTERNATING CURRENT

10.1 ALTERNATING QUANTITIES

As mentioned earlier, an alternating quantity is one which reverses its direction periodically, being in one direction (say the 'positive' direction) at one moment and in the opposite direction (the 'negative' direction) the next moment. The frequency of alternations can be as low as once every few seconds or as high as once every few nanoseconds (1 ns = 10^{-9} s or $\frac{1}{1\,000\,000\,000}$ s).

The **frequency** of the UK alternating power supply is 50 cycles per second or 50 hertz (Hz), so that the time for one complete cycle (known as the **periodic time** of the supply) is

$$T = \frac{1}{f} \text{ seconds} = \frac{1}{50} = 0.02 \text{ s or } 20 \text{ ms}$$

Whilst many countries have adopted 50 Hz as the supply frequency, other countries such as the US, use a frequency of 60 Hz, having a periodic time of

$$T = \frac{1}{60} \text{ s} = 0.01667 \text{ s or } 16.67 \text{ ms}$$

Radio transmissions use a much higher frequency, and a frequency of 10 MHz has a periodic time of

$$\frac{1}{(10 \times 10^6)} = 0.1 \times 10^{-6} \text{ s or } 0.1 \text{ } \mu\text{s}$$

The electricity power supply has a sinusoidal waveform of the type in Figure 10.1. The waveform shown is that of a current wave which follows the equation

$$i = I_m \sin \theta \qquad (10.1)$$

fig 10.1   *a sinusoidal current waveform*

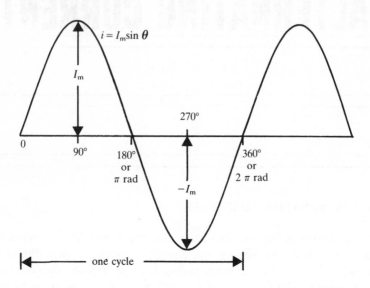

where i is the instantaneous value of the current at angle θ, and I_m is the **maximum value** or **peak value**, which is reached at an angle of 90°. The current follows the sinewave through zero, and then increases to its maximum negative value of $-I_m$ at an angle of 270°. The cycle repeats itself every 360°. As mentioned in Chapter 8, the *angular frequency* of a wave is given in radians per second, where

$$\text{angular frequency, } \omega = 2\pi \text{ rad/s} \qquad (10.2)$$

so that the angular frequency corresponding to a frequency of 50 Hz is

$$\text{angular frequency, } \omega = 2\pi \times 50 = 100\pi = 314.2 \text{ rad/s}$$

It was shown in Chapter 8 that an angle in radians is given by

$$\text{angle in radians} = \text{angle in degrees} \times \frac{2\pi}{360}$$

so that

$$\text{for } \theta = 90°, \text{ radian angle} = 90 \times \frac{2\pi}{360} = \frac{\pi}{2} \text{ rad}$$

$$\text{for } \theta = 180°, \text{ radian angle} = 180 \times \frac{2\pi}{360} = \pi \text{ rad}$$

A complete cycle ($360°$) is therefore equivalent to

$$\text{radian angle} = 360 \times \frac{2\pi}{360} = 2\pi \text{ rad}$$

If the angular frequency of the waveform is ω rad/sec, then the radian angle, θ, turned through after a time t seconds is

$$\text{radian angle, } \theta = \text{angular velocity (rad/s)} \times \text{time (s)}$$

$$= \omega t \text{ rad}$$

If the supply frequency is 50 Hz (or $\omega = 2\pi f = 100 \, \pi$ rad/s), then the angle turned through when $t = 5$ ms after the start of the sinewave is

$$\theta = \omega t = 100\pi \times (5 \times 10^{-3}) = \frac{\pi}{2} \text{ rad (or } 90°)$$

The above equation allows eqn (10.1) to be rewritten in the form

$$\text{current, } i = I_m \sin \theta = I_m \sin \omega t$$

So far we have discussed only a current wave. If the sine wave is that of a voltage, the equation can be written as follows

$$\text{voltage, } v = V_m \sin \theta = V_m \sin \omega t \qquad (10.3)$$

where v is the *instantaneous voltage* of the wave at angle θ after the commencement of the wave (or at time t after the start of the wave), V_m is the *maximum value* or *peak value* of the wave, and ω is the angular frequency of the wave.

10.2 MEAN VALUE OR AVERAGE VALUE OF A SINE WAVE

The strict meaning of the average value of a waveform is

$$\text{average value} = \frac{\text{total area under one complete wave}}{\text{periodic time of the wave}}$$

However, an alternating waveform *has equal positive and negative areas*, so that the **total area under the wave taken over a complete cycle is zero**. That is, the average value of an alternating wave taken over one complete cycle is zero!

The *electrical engineering interpretation of the average value* or *mean value* therefore differs from this and is given by

$$\text{mean value} = \frac{\text{area under one half of the waveform}}{\text{one-half of the periodic time}} \qquad (10.4)$$

To determine the area under the half cycle, we shall use the *mid-ordinate rule* as follows. The half cycle (a current waveform is chosen in this case – see Figure 10.2) is divided into an equal number of parts by lines known as *ordinates* (shown as dotted lines in the figure). *Mid-ordinates i_1, i_2, i_3,* etc, are drawn and measured, and the average value of the current in that half cycle is calculated as follows:

$$\text{average current, } I_{av} = \frac{\text{sum of the mid-ordinates}}{\text{number of mid-ordinates}} \qquad (10.5)$$

In Figure 10.2 only three mid-ordinates are shown, which is far too few to give a reliable answer; many more values are usually needed to give an accurate result.

fig 10.2 *the 'mean' value or 'average' value of an a.c. wave*

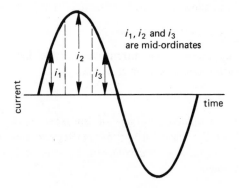

i_1, i_2 and i_3 are mid-ordinates

Suppose that the current waveform has a maximum value of 1 A and that we can divide the waveform up to give ten $18°$ mid-ordinates (the first being at $9°$ and the last at $171°$). If the values of the mid-ordinates are as listed in Table 10.1, the average value of the current is

$$\text{average current, } I_{av} = \frac{\text{sum of mid-ordinate values}}{\text{number of mid-ordinates}}$$

$$= \frac{6.3922}{10} = 0.63922 \text{ A}$$

The above average value was, in fact, calculated for a sine wave using only ten mid-ordinates. A more accurate value is determined using more mid-ordinates, the most accurate value being obtained using the **calculus** which can be thought of as taking an infinite number of mid-ordinates. It can be shown that using the calculus the **mean value of a sinewave of current** is

$$\text{mean current, } I_{av} = 0.637 I_m \qquad (10.6)$$

where I_m is the maximum value of the current. In our case, the maximum value of the current is 1 A, so that the calculus gives an average current of 0.637 A (compared with our value of 0.639 A).

The same relationship holds for a voltage sine wave; that is **the average value of a sinusoidal voltage wave** is

$$\text{mean voltage, } V_{av} = 0.637\, V_m \qquad\qquad (10.7)$$

Table 10.1 *Calculation of average value of current*

Angle	Mid-ordinate current
9°	0.1564
27°	0.4539
45°	0.7071
63°	0.891
81°	0.9877
99°	0.9877
117°	0.981
135°	0.7071
153°	0.4539
171°	0.1564

sum of mid-ordinates = 6.3922 A

Example
Calculate the mean value of a sinusoidal voltage wave whose maximum value is 100 V. Determine also the maximum value of a voltage wave whose average value is 90 V.

Solution
(i) $V_m = 100$ V

$$\text{mean voltage, } V_{av} = 0.637 V_m = 0.637 \times 100$$
$$= 63.7 \text{ V (Ans.)}$$

(ii) $V_{av} = 90$ V

$$\text{maximum value, } V_m = \frac{V_{av}}{0.637} = \frac{90}{0.637} = 141.3 \text{ V (Ans)}$$

10.3 THE EFFECTIVE VALUE OR ROOT-MEAN-SQUARE (r.m.s.) VALUE OF A SINE WAVE

The **effective value** of an alternating current (or an alternating voltage for that matter) is expressed in terms of its heating effect; that is, it is given in terms of its I^2R effect. The effective value is known as the **root-mean-square** (r.m.s.) value of the wave and is calculated as follows.

> r.m.s. value = **square root** of the **mean** of the sum of **squares** (r.m.s.) of the mid-ordinate values of the wave.

This is illustrated in Figure 10.3. Each value on the waveform (the current waveform in this case) is multiplied by itself (that is, it is 'squared'); if the wave is a sinewave, the equation of the wave is $I_m \sin \theta$, and the equation of the (current)^2 graph is $I_m^2 \sin^2\theta$. You will see that the (current)^2 graph has a positive value in the second half cycle even though the current is negative (this is because the product of two negative values is a positive value). The r.m.s. value of a sinusoidal current wave is given by

$$\text{r.m.s. current, } I = \sqrt{\frac{(\text{sum of the mid-ordinate (current)}^2 \text{ values})}{\text{number of mid-ordinates}}}$$

(10.8)

or, alternatively

$$\text{r.m.s. current, } I = \sqrt{\frac{(\text{area under (current)}^2 \text{ curve})}{(\text{length of base of the waveform})}}$$

(10.9)

fig 10.3 *the r.m.s. value of an alternating wave*

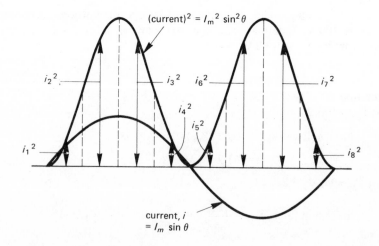

In the following we will determine the r.m.s. value of a sinusoidal current waveform whose values are given in Table 10.2; mid-ordinates are taken at $36°$ intervals (the first at $18°$). The results in the table give

sum of $(current)^2 = 5.0$ $(amperes)^2$

hence

$$r.m.s. \text{ value of current} = \sqrt{\frac{\text{sum of }(current)^2}{(\text{number of mid-ordinates})}}$$

$$= \sqrt{\frac{5}{10}} = 0.7071 \text{ A}$$

That is to say, for a *current sine wave* the r.m.s. value is given by

$$r.m.s. \text{ current}, I = 0.7071 I_m = \frac{I_m}{\sqrt{2}} \tag{10.10}$$

Similarly, for a *sinusoidal voltage* the r.m.s. value is given by

$$r.m.s. \text{ voltage}, V = 0.7071 V_m = \frac{V_m}{\sqrt{2}} \tag{10.11}$$

Example
(i) Calculate the r.m.s. value of a sinusoidal current wave of maximum value 20 A. Determine (ii) the maximum value of a voltage wave whose r.m.s. value is 240 V.

Table 10.2 *Calculation of r.m.s. value of current*

Angle	Value of current	Value of $(current)^2$
$18°$	0.3090	0.0955
$54°$	0.8090	0.6545
$90°$	1.0	1.0
$126°$	0.8090	0.6545
$162°$	0.3090	0.0955
$198°$	-0.3090	0.0955
$234°$	-0.8090	0.6545
$270°$	-1.0	1.0
$306°$	-0.8090	0.6545
$342°$	-0.3090	0.0955
	sum of $(current)^2 =$	5.0

Solution

(i) I_m = 20 A

r.m.s. current, I = 0.7071 I_m = 0.7071 × 20

= 14.14 A (Ans.)

(ii) V_S = 240 V r.m.s.

From eqn (10.7), $V_S = \frac{Vm}{\sqrt{2}}$, hence

maximum voltage, $V_m = V_S \times \sqrt{2}$ = 240 × 1.414

= 339.36 V (Ans.)

10.4 AVERAGE VALUE AND r.m.s. VALUE OF A WAVE OF ANY SHAPE

So far we have discussed the mean and r.m.s. values of a sinewave. Every waveform has its own average and r.m.s. value, and can be calculated for current waves from eqns (10.4) and (10.5) and for voltage waveforms from eqns (10.8) and (10.9). Illustrative calculations are performed for the triangular wave in Figure 10.4.

Since the waveform has equal positive and negative areas, the mean value must be calculated over one half cycle; for convenience, the positive half-cycle is used. In this case the values are for a voltage wave (but it could well be a current wave!) and, for the wave in Figure 10.4:

mean voltage, V_{av} = $\dfrac{\text{sum of mid-ordinate voltages}}{\text{number of mid-ordinates}}$

$$= \frac{(0.25 + 0.75)}{2} = 0.5 \text{ V (Ans.)}$$

The r.m.s. value of the wave in Figure 10.4 is calculated over a complete cycle using the equation

r.m.s. voltage, $V = \sqrt{\dfrac{\text{(sum of mid-ordinate (current)}^2 \text{ values)}}{\text{(number of mid-ordinates)}}}$

$$= \sqrt{\left(\frac{([-0.75]^2 + [-0.25]^2 + 0.25^2 + 0.75^2)}{4} \right)}$$

= 0.559 V (Ans.)

Note

You should be aware of the fact that taking only a few mid-ordinates may give a result of low accuracy. Whilst the mean voltage calculated above is

fig 10.4 *mean and r.m.s. values of a non-sinusoidal wave*

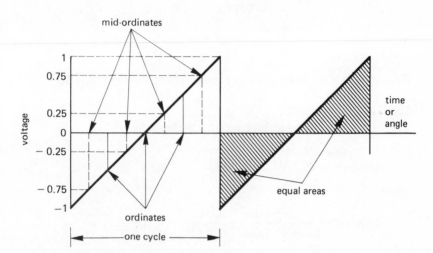

perfectly correct, the r.m.s. value obtained above is of limited accuracy (the value using many mid-ordinates is 0.5774 V.)

10.5 FORM FACTOR AND PEAK FACTOR OF A WAVEFORM

Information relating to the 'shape' of an a.c. waveform is often useful to electrical engineers. The factors known as the **form factor** and the **peak factor** (the latter also being known as the **crest factor**) act as electrical 'fingerprints' of the wave. Two differing waveforms may have the same value for one of the factors, but the other factor will differ between the two waves, indicating that the waveforms are different in shape. The two factors are defined below

$$\text{form factor} = \frac{\text{r.m.s. value of wave}}{\text{mean value of wave}} = \frac{I}{I_{av}} \text{ or } \frac{V}{V_{av}} \qquad (10.12)$$

and

$$\text{peak factor} = \frac{\text{peak value of wave}}{\text{r.m.s. value of wave}} = \frac{I_m}{I} \text{ or } \frac{V_m}{V} \qquad (10.13)$$

The **form factor and peak factor for a sinewave** (taking a current wave in this instance) are calculated as follows:

$$\text{form factor} = \frac{I}{I_{av}} = \frac{0.7071 \, I_m}{0.637 \, I_m}$$

$$= 1.11$$

$$\text{peak factor} = \frac{I_m}{I} = \frac{I_m}{0.7071\,I_m}$$
$$= 1.414$$

Example

Calculate the form factor of the triangular voltage wave in Figure 10.4.

Solution

The r.m.s. and mean values estimated for the wave are

r.m.s. value = 0.559 V
mean value = 0.5 V

hence

$$\text{form factor} = \frac{\text{r.m.s. value}}{\text{mean value}} = \frac{0.559V}{0.5V} = 1.118$$

Note

The above figure differs only slightly from that of the sinewave, the reason being that the r.m.s. value determined in the calculation is not sufficiently accurate. If the true r.m.s. value of 0.5774 V is used, the form factor of the wave is seen to be

$$\frac{0.5774}{0.5} = 1.155$$

10.6 PHASE ANGLE DIFFERENCE BETWEEN TWO SINEWAVES

Suppose that we have a two-coil alternator on which both coils have the same number of turns but are in different physical positions on the rotor. Because of the difference in the position of the coils, the e.m.f. in each coil will differ, as illustrated in the waveforms in Figure 10.5 for coils A and B on the rotor of the alternator.

As each coil rotates, a sinewave of voltage is induced in it, but each wave differs from the other by an angle ϕ which is known as the *phase angle* difference (which can be expressed either in degrees or radians). The phase angle difference (often simply referred to as the *phase angle*) is the *angular difference between the two waves when they are at the same point on their waveform*. For example, the phase angle difference can be measured as the angular difference between the waves when they both pass through zero and are increasing in a positive direction (see Figure 10.5); alternatively it is the angular difference between the two waves when they pass through their peak negative voltage (also see Figure 10.5).

fig 10.5 *phase angle difference*

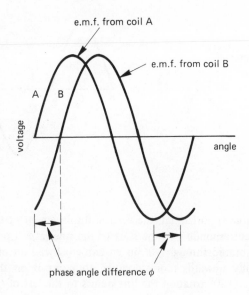

Lag or lead?

When describing phase angle difference, it is frequently necessary to know which wave 'lags' or which wave 'leads' the other, that is, which wave is 'last'and which is 'first'? The answer to this is revealed from a study of Figure 10.5. Clearly, waveform A passes through zero *before* waveform B; from this we can say that

> **waveform A leads waveform By by angle** ϕ

Alternatively, since waveform B passes through zero *after* waveform A, we may say that

> **waveform B lags behind waveform A by angle** ϕ

Both statements are equally correct.

10.7 PHASOR DIAGRAMS

Waveform diagrams are difficult to visualise, and engineers have devised a diagrammatic method known as a **phasor diagram** to simplify the problem.

Imagine a line of length V_m rotating in an anti-clockwise direction (see Figure 10.6(a)). If you plot the *vertical displacement* of the tip of the line at various angular intervals, the curve traced out is a sinewave (see Figure 10.6(b)).

fig 10.6 *production of a sinewave*

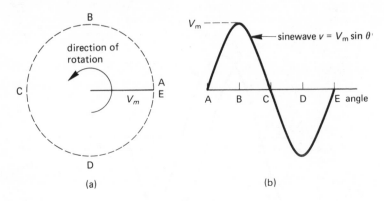

(a)

(b)

When the line is horizontal, the vertical displacement of the tip of the line is zero, corresponding to the start of the sinewave at point A. After the line has rotated through 90° in an anti-clockwise direction, the line points vertically upwards (corresponding to point B on the waveform diagram). After 180° rotation the line points to the left of the page, and the vertical displacement is zero once more (corresponding to point C on the waveform diagram). After a further 180° (360° in all) the rotating line reaches its starting position once more (corresponding to point E on the waveform diagram).

A **phasor** is a line representing the rotating line V_m, but is **scaled to represent the r.m.s. voltage, V, which is 'frozen' at some point in time.**

For example, if the rotating line is 'frozen' at point A in Figure 10.6(b) the corresponding voltage phasor which represents this is shown in Figure 10.7(a). If the rotating line is 'frozen' at point B, the voltage phasor is as shown in Figure 10.7(b). Voltage phasors representing the 'freezing' of the rotating line at points C, D and E in Figure 10.6 are represented by the

fig 10.7 *several phasor diagrams for Figure 10.5(b)*

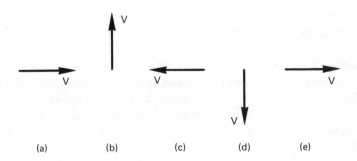

(a)   (b)   (c)   (d)   (e)

voltage phasors in diagrams (c), (d) and (e) in Figure 10.7. You will see that the 'length' of the phasor in Figure 10.7 is 0.7071 that of the rotating line in Figure 10.5 (remember, r.m.s. voltage = 0.7071 × maximum voltage).

Phasor representation of two waveforms

Consider the case of the two waveforms in Figure 10.8, one being a voltage wave and the other a current wave, which are out of phase with one another (in this case the current waveform lags behind the voltage waveform by angle ϕ).

If the phasor diagram for the two waves is drawn corresponding to point W on the wave, the corresponding phasor diagram is shown in diagram (a) in Figure 10.8 (**note**: the current phasor lags behind the voltage phasor in the direction of 'rotation' of the phasor).

fig 10.8 *phasor diagrams for a voltage wave and a current wave which are out of phase with one another*

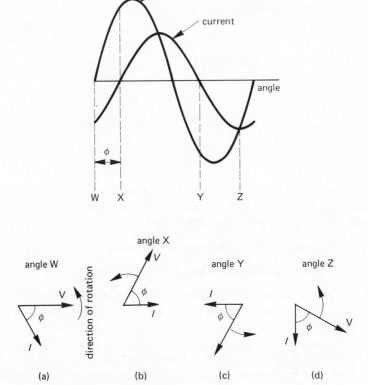

If, on the other hand, the wave is 'frozen' at point X, the corresponding phasor diagram is shown in diagram (b). Once again, the current phasor 'lags' behind the voltage phasor.

The phasor diagrams corresponding to points Y and Z on the waveform diagram are drawn in diagrams (c) and (d), respectively; you will note in each case that the current phasor always lags behind the voltage phasor.

10.8 ADDITION OF PHASORS

When two alternating voltages are connected in series in an a.c. circuit as shown in Figure 10.9, the two voltage waveforms may not be in phase with one another. That is, **you need to take account of the fact that there is a phase angle difference between the voltages**. It is rather like the case of a tug-of-war team in which one of the team pulls in the wrong direction; the total pull is not simply the sum of the pulls of the individual team members; account must be taken of the 'direction' of the pulls. In the a.c. circuit you must determine the **phasor sum** of the voltages.

fig 10.9 *phasor addition of voltages*

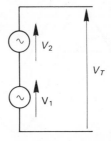

Suppose that voltage V_1 has an r.m.s. voltage of 15 V and V_2 has an r.m.s. voltage of 10 V, but V_2 leads V_1 by 60° (see Figure 10.10). The total voltage, V_T, in the circuit is given by

total voltage, V_T = **phasor sum** of V_1 and V_2

The solution can be obtained in one of two ways, namely:

1. the phasor diagram can be drawn to scale and the magnitude and phase angle of the resultant voltage, V_T, can be determined from the scale diagram;
2. you can 'add' the voltages together by resolving the components of the two voltages in the 'vertical' and 'horizontal' directions, and from this determine the 'vertical' and 'horizontal' components of the voltage V_T;

the magnitude and phase angle of V_T can be determined from the resolved components of V_T.

Method 1 needs only simple drawing instruments and, provided that the scale selected for the drawing is large enough, the method gives an answer which is accurate enough for most purposes. Figure 10.10 shows how this is done for the voltages given above. You will find it an interesting exercise to draw the phasor diagram to scale and to compare your results with the answer calculated below.

The solution using method 2 is described below for the voltages given above (see also the phasor diagram in Figure 10.10). First, the voltages V_1 and V_2 are resolved into their horizontal and vertical components.

Voltage V_1

Since we are starting with this voltage, we can assume that it points in the horizontal or 'reference' direction. That is to say, the voltage has a horizontal component and no vertical component, as follows:

Horizontal component = 15 V
Vertical component = 0 V

Voltage V_2

This voltage leads V_1 by an angle which is less than 90°, so that it has both horizontal and vertical components, which are calcualted as shown:

Horizontal component = $10 \cos 60° = 10 \times 0.5 = 5$ V

Vertical component = $10 \sin 60° = 10 \times 0.866 = 8.66$ V

Resultant voltage, V_T

The horizontal and vertical components are calculated as follows:

horizontal component = sum of horizontal components of V_1 and V_2

$$= 15 + 5 = 20 \text{ V}$$

fig 10.10 *the phasor sum of two voltages*

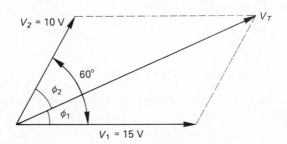

$$\text{vertical component} \quad = \text{sum of the vertical components of } V_1 \text{ and } V_2$$

$$= 0 + 8.66 = 8.66 \text{ V}$$

The magnitude of voltage V_T is determined by Pythagorus's theorem as follows (see also Figure 10.11)

$$V_T = \sqrt{[(\text{horizontal component of } V_T)^2 + (\text{vertical component of } V_T)^2]}$$

$$= \sqrt{(20^2 + 8.66^2)} = \sqrt{475}$$

$$= 24.8 \text{ V (Ans.)}$$

fig 10.11 *determination of the voltage V_T*

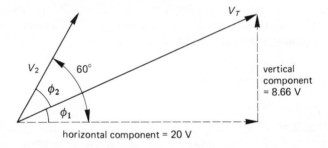

horizontal component = 20 V

The value of the phase angle ϕ between V_T and V_1 is calculated from the equation

$$\tan \phi = \frac{8.66}{24.8} = 0.349$$

hence

$$\phi_1 = \tan^{-1} 0.347 = 19.25°$$

where 'tan^{-1}' means 'the angle whose tangent is'. You may therefore say that V_T leads V_1 by 19.25°. However

$$\phi_1 + \phi_2 = 60°$$

or

$$\phi_2 = 60° - \phi_1 = 60° - 19.25° = 40.75°$$

where ϕ_2 is the angle between V_2 and V_T. It follows that V_T lags behind V_2 by 40.75°.

10.9 VOLT–AMPERES, WATTS AND VOLT–AMPERES REACTIVE

In a d.c. circuit, the product 'volts × amperes' gives the power consumed by the circuit. The situation is slightly more complex in an a.c. circuit, and we will study this situation in this chapter.

Let us look at a circuit in which the voltage and current waveforms are in phase with one another, that is, a circuit containing only a pure resistor, as shown in Figure 10.12(a). When the two waves are multiplied together to give the volt-ampere product, as shown in Figure 10.12(b), the power consumed by the circuit is given by the *average value of the area which is between the zero line and the volt-ampere graph*. You can see that, in this case, power is consumed all the time that current flows in the circuit.

fig 10.12   *(a) waveform diagram for a sinusoidal voltage wave and a current wave which are in phase with one another and (b) the corresponding volt–ampere product wave*

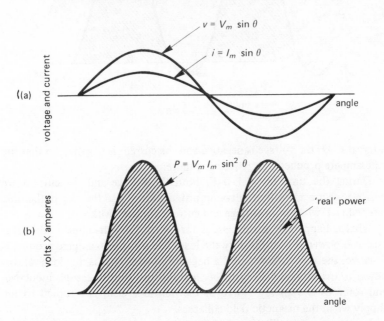

Let us now consider a circuit in which the current lags behind the voltage by an angle of 45° (see Figure 10.13(a)). When the voltage and current waveforms are multiplied together (Figure 10.13(b)) we see that in the time interval between A and B the voltage is positive and the current is negative; the volt-ampere product is therefore negative. Also in the time

fig 10.13 *waveforms of voltage, current and volt-ampere product for two waves which are 45° out of phase with one another*

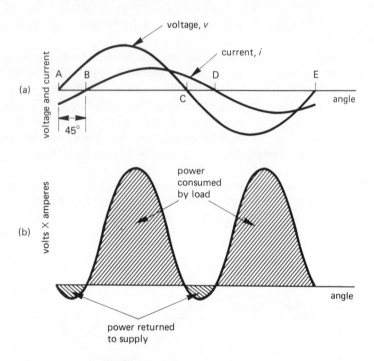

interval C-D, the voltage is negative and the current is negative, so that the volt-ampere product is negative once more.

During the time interval B-C, both the voltage and the current are positive, giving a positive volt-amp product (as is also the case in the time interval D-E when both voltage and current are negative).

What is happening in this case is that when the volt-ampere product is positive, power is consumed by the load. When the volt-ampere product is negative, the power 'consumes' a negative power! That is, the load returns power to the supply; this occurs because the load (in this case) is inductive, and some of the energy stored in the magnetic field is returned to the supply when the magnetic field collapses.

You can see from Figure 10.13(b) that more energy is consumed by the load than is returned to the supply and, on average over the complete cycle, power is consumed by the load.

We turn our attention now to the case where the phase angle between the voltage and current is 90° (see Figure 10.14). In this case, the volt-ampere product graph has equal positive and negative areas, which tells us that **the load returns as much power to the supply as it consumes**. That is,

fig 10.14 *waveforms for a phase angle of 90°; the average value of the power consumed is zero*

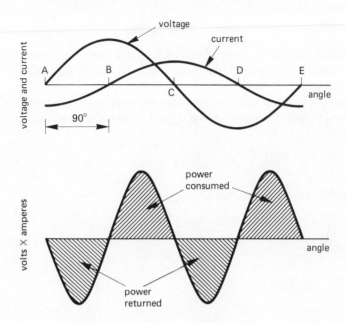

when the phase angle between the voltage and the current is 90°, the average power consumed by the load is zero.

It follows from the above that *the volt–ampere product in an a.c. circuit does not necessarily give the power consumed.* In order to determine the power consumed by an a.c. circuit, you need to account not only for the volt-ampere product but also for the phase angle between the voltage and current.

The product of the voltage applied to an a.c. circuit and the current in the circuit is simply known as the **volt-ampere product** (shortened to VA by engineers), and is given the symbol S. If the supply voltage is V_S and the current taken by the circuit, the VA consumed is

$$\text{volt-amperes, } S = V_S I \text{ VA} \tag{10.14}$$

The volt-ampere product is also known as the **apparent power** consumed by the circuit or device.

The 'real' **power** or useful power in watts used to produce either work or heat is given by the equation

$$\text{power, } P = V_S I \cos \phi \text{ W} \tag{10.15}$$

where ϕ is the phase angle between V_S and I.

There is also another element to the power 'triangle' in a.c. circuits which is the **reactive power** or **volt-amperes reactive** (VAr) symbol Q, and is given by the equation

reactive volt-amperes, $Q = V_S I \sin \phi \ VAr$ (10.16)

The three elements of 'power' in an a.c. circuit can be represented by the three sides of the **power triangle** in Figure 10.15 (for further details see Chapter 11).

Since the three 'sides' representing the apparent power (S), the real power (P), and the reactive power (Q) are related in a right-angled triangle, then

$$S^2 = P^2 + Q^2 \tag{10.17}$$

fig 10.15 *the power triangle of an a.c. circuit*

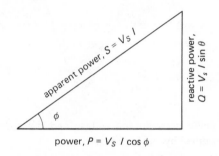

power, $P = V_S I \cos \phi$

Example

An a.c. circuit supplied at 11000 V r.m.s. draws a current of 50 A. The phase angle between the applied voltage and the current is 36°. Calculate for the circuit (i) the apparent power, (ii) the 'real' power and (iii) the reactive power consumed.

Solution

$V_S = 11\,000$ V; $I = 50$ A; $\phi = 36°$

(i) apparent power, $S = V_S I = 11\,000 \times 50 = 550\,000$ VA (Ans.)

(ii) power, $P = V_S I \cos \phi = 11\,000 \times 50 \times \cos 36°$

$= 444\,959$ W or 444.959 kW (Ans.)

(iii) reactive power, $Q = V_S I \sin \phi = 11\,000 \times 50 \times \sin 36°$

$= 323\,282$ VAr or 323.282 kVAr

You should note that whilst the circuit consumes 550 kVA (volts × amps), the useful power output from the circuit (which may be, for example, heat) is about 445 kW. The relationship between these two quantities is discussed further in section 10.10.

10.10 POWER FACTOR

The **power factor** of an a.c. circuit is the **ratio of the useful power (in watts [W]) consumed by a circuit to the apparent power (VA) consumed**, and is given by the equation

$$\text{power factor} = \frac{\text{'real' power in watts}}{\text{apparent power in volt-amperes}}$$

$$= \frac{V_S I \cos \phi}{V_S I} = \cos \phi \qquad (10.18)$$

where ϕ is the phase angle between V_S and I.

If for example, the phase angle between the current and the voltage is 0°, the power factor of the circuit is

power factor = $\cos 0° = 1.0$

That is, the 'real' power is equal to the apparent power, that is the power in watts consumed is equal to the number of volt–amperes consumed. Waveforms for this condition are shown in Figure 10.12, and correspond to the example of an a.c. circuit containing a pure resistive load.

If the phase angle between the voltage and the current is 45°, the power factor of the circuit is

power factor = $\cos 45° = 0.7071$

That is, for each 100 VA consumed by the circuit, 70.71 W are consumed. The reason why not all the VA is converted into watts is shown in the waveform in Figure 10.13.

If the phase angle of the circuit is 90°, the circuit power factor is

power factor = $\cos 90° = 0$

In this case the power consumed by the circuit is zero even though the current may be very large. The reason for this can be seen from the waveforms in Figure 10.14, where it was shown that the circuit returns as much power to the supply as it takes from it. This situation arises in any circuit where the phase angle between the voltage and the current is 90°, that is, in either a circuit containing a pure inductor or a pure capacitor.

Equation (10.18) may be re-written in the following form:

Power = apparent power × power factor

$$= V_S I \cos \phi \text{ W} \tag{10.19}$$

Example
If the supply voltage to a circuit of power factor 0.8 is 11 000 V, and the power consumed in 200 kW, calculate the current in the circuit.

Solution
V_S = 11 000 V; P = 200 000 W; power factor = 0.8
From eqn (10.19) $P = V_S I \cos \phi$, hence

$$I = \frac{P}{(V_S \cos \phi)} = \frac{200\,000}{(11\,000 \times 0.8)}$$

$$= 22.73 \text{ A (Ans.)}$$

10.11 HARMONICS IN a.c. SYSTEMS

A device such as a resitor has a characteristic in which the current through it is proportional to the voltage applied to it (see Figure 10.16(a)). This type of characteristic is known as a **linear characteristic** or *straight line characteristics*; if the voltage applied to it is a sinewave, the current through it is also a sinewave.

However, in a device which has a voltage–current characteristic which is not a straight line such as that in Figure 10.16(b), the current flowing through it is not proportional to the applied voltage. That is, if a sinusoidal voltage is applied, the current flowing through it is non-sinusoidal. This type of characteristic is said to be **non-linear**, that is, it is not a straight line which passes through zero.

Many common electrical devices have a non-linear characteristic, including fluorescent lights and iron-cored inductors. This results in the current waveshape being non-sinusoidal even though the voltage waveshape is sinusoidal; these current waveshapes are said to be **complex**.

Complex waveshapes which arise in circuits can be constructed or **synthesised** by adding together a series of sinewaves known as **harmonics** as follows:

Fundamental frequency
The fundamental frequency is a sinewave whish is the 'base' frequency on which the complex wave is built. The periodic time of the complex wave is equal to the periodic time of the fundamental frequency.

fig 10.16 *(a) a linear characteristic and (b) a non-linear characteristic*

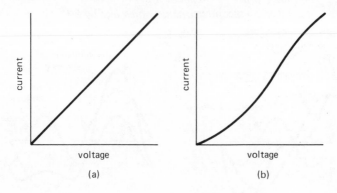

(a)   (b)

Second harmonic
This is a sinewave whose frequency is twice that of the fundamental frequency;

Third harmonic
This is a sinewave whose frequency is three times that of the fundamental frequency;

Fourth harmonic
This is a sinewave whose frequency is four times that of the fundamental frequency;

100th harmonic
This is a sinewave whose frequency is 100 times that of the fundamental etc.

The 'shape' of the resulting complex wave depends not only on the number and amplitude of the harmonic frequencies involved, but also on the phase relationship between the fundamental and the harmonics, as illustrated in Figure 10.17 which shows a fundamental frequency together with a second harmonic.

In Figure 10.17(a), the second harmonic has one-half the amplitude of the fundamental, and commences in phase with the fundamental. The complex wave produced by the addition of the two waves is symmetrical about the 180° point of the wave, the wave being 'peaky' in both half cycles.

In Figure 10.17(b), the same second harmonic is added to the fundamental but, in this case, the second harmonic lags behind the fundamental

fig 10.17   *complex wave formed from (a) a fundamental and a second harmonic which are in phase with one another, (b) a fundamental and a second harmonic which lags by 90°*

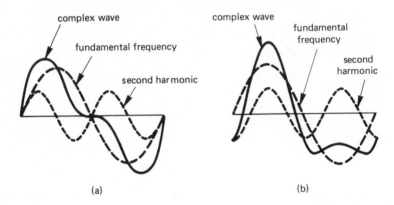

(a)   (b)

frequency by 90° of the harmonic wave; the resulting complex wave is peaky in the first half cycle but has a flattened second half cycle.

If high-frequency harmonics are present in a wave, then the sound of the resultant complex wave is 'sharp' to the ear, that is it has a treble sound. However, the phase angle between the fundamental frequency and the harmonics has very little effect so far as hearing the sound is concerned since the human ear is insensitive to phase shift. Unfortunately, if the harmonic is produced in a TV video circuit, the phase shift of the harmonic produces quite a marked effect on the TV screen since it gives rise to a change in colour (we are all familiar with coloured 'interference' band patterns on the TV tube produced under certain conditions which result from harmonic effects).

Power electronic devices such as thyristors also give rise to harmonics in the power supply system. In industry, these harmonic currents are frequently carried by overhead power lines which can act as an aerial, giving rise to radiated electromagnetic energy; the radiated energy can produce interference in radios, TVs and other electronic equipment. It is for this reason that electricity supply authorities limit the amount of harmonic current that may be produced by industry.

SELF-TEST QUESTIONS

1. Calculate the periodic time of an alternating wave of frequency 152 Hz. Determine also the frequency of an alternating wave whose periodic time is 0.25 μs.

2. A sinusoidal current wave has a value of 10 A when the angle of rotation is 58°. Calculate (i) the peak-value of the current and (ii) the value of the current at an angle of 10°, 120°, 200°, 350°, 1 rad, 2 rad, and 4 rads.

3. Calculate the mean value and the r.m.s. value of a voltage sinewave of peak value 120 V.

4. What is meant by the 'form factor' and 'peak factor' of an alternating wave? Why do different types of wave have different combinations of form factor and peak factor?

5. Two series-connected sinusoidal voltages of r.m.s. value 200 V and 150 V, respectively, the 200-V wave leading the 100-V wave by 45°. Determine (i) the sum of, and (ii) the difference between, the waveforms.

6. An a.c. circuit consumes 10 kW of power when the applied voltage is 250 V. If the current is 50 A, determine (i) the VA consumed, (ii) the phase angle and the power factor of the circuit and (iii) the VAr consumed.

SUMMARY OF IMPORTANT FACTS

An **alternating quantity** is one which periodically reverses its direction or polarity. The **frequency**, f, of the wave is the number of alternations per second (Hz), and the **periodic time**, T, is the time taken for one complete cycle of the wave ($f = \frac{1}{T}$).

The **mean value** of a wave is the **average value** taken over a given period of the wave (in the case of an alternating quantity, the period is *one-half* of the periodic time of the wave). The **effective value** or **root-mean-square** (r.m.s.) value of the wave is the *effective heating value* of the wave. For a *sinewave*

mean value = 0.637 × maximum value

r.m.s. value = 0.7071 × maximum value

The factors which give some indication of the 'shape' of the waveform are the **form factor** and the **peak factor**. For a *sinewave* these factors are

form factor = 1.11
peak factor = 1.414

The angular difference between two sinewaves is known as the **phase difference**. A **phasor diagram** represents waveforms which are *frozen in time*, and are scaled to represent the r.m.s. values of the waves concerned. Phasors can be added to or subtracted from one another to give resultant values.

The product (volts x amperes) in an a.c. circuit is known as the **volt-ampere product (VA)** or **apparent power**. The **real power** or heating effect is measured in watts. The **reactive volt-amperes (VAr)** or **reactive power** is the volt-ampere product which does not produce any 'real' power consumption in the circuit.

The **power factor** of a circuit is given by the ratio of watts: volt-amperes.

Non-sinusoidal waveforms in a circuit can be thought of as consisting of a series of sinewaves known as **harmonics**, which are added together to form the complex wave in the circuit.

INTRODUCTION TO

SINGLE-PHASE a.c.

CIRCUITS

11.1 RESISTANCE IN AN a.c. CIRCUIT

In an a.c. circuit at normal power frequency, a resistance behaves in the same way as it does in a d.c. circuit. That is, *any change in voltage across the resistor produces a proportional change in current through the resistor* (Ohm's law).

Suppose that a pure resistor is connected to a sinusoidal voltage as shown in Figure 11.1(a). The current through the resistor varies in proportion to the voltage, that is **the current follows a sinewave which is in phase with the voltage** (see Figure 11.1(b)). The corresponding phasor diagram for the circuit is shown in Figure 11.1(c).

Since *the phase angle between the voltage and the current is zero*, then the *power factor* of the resistive circuit is cos 0° = 1.0. The r.m.s. value of the current, I, in the circuit is calculated from Ohm's law as follows:

$$\text{current}, I = \frac{V_R}{R} \tag{11.1}$$

fig 11.1 *a pure resistor in an a.c. circuit*

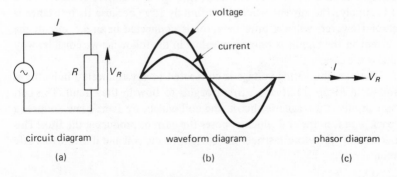

circuit diagram
(a)

waveform diagram
(b)

phasor diagram
(c)

where V_R is the r.m.s. value of the voltage across the resistor. The power, P, consumed in the circuit is given by

power, $P = V_R I \cos \phi$ W

but, since $\cos \phi = 1$ in a resistive circuit, then

$P = V_R I$ W

However, by Ohm's law, $V_R = IR$, then

power, $P = V_R I = (IR) \times I = I^2 R$ W $\hspace{2cm}$ (11.2)

Example

Calculate the power consumed in a single phase a.c. circuit which is energised by a 240-V, 50-Hz sinusoidal supply, the load being a resistance of value 100 Ω.

Solution

$\hspace{1cm} V_S = 240$ V; $R = 100$ Ω

Since the only element in the circuit is the resistor, then

$\hspace{1cm} V_R = V_S = 100$ V

From eqn (11.1)

$\hspace{1cm}$ current, $I = \dfrac{V_R}{R} = \dfrac{240}{100} = 2.4$ A

and from eqn. (11.2)

$\hspace{1cm}$ power consumed, $P = I^2 R = 2.4^2 \times 100 = 576$ W (Ans.)

11.2 PURE INDUCTANCE IN AN a.c. CIRCUIT

A pure inductance is resistanceless, and its only property is that it produces a magnetic flux when a current flows through it. If it were connected to a d.c. supply, the current would be infinitely large because its resistance is zero! However, when a pure inductor is connected in an a.c. circuit the current in the circuit is limited in value; in the following we consider why this is the case.

When an alternating voltage is connected to a resistanceless inductor as shown in Figure 11.2(a), a current begins to flow in the circuit. This current produces a magnetic flux in the coil which, by Lenz's law, induces a 'back' e.m.f. in the coil which opposes the current producing the flux. This back e.m.f. therefore restricts the current in the coil and limits it to a safe value.

fig 11.2 *a pure inductor in an a.c. circuit*

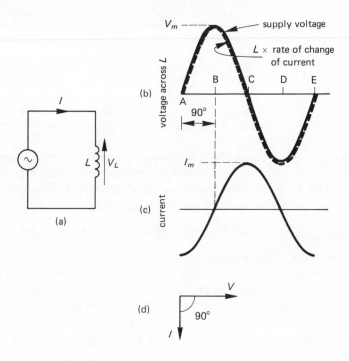

This phenomenon gives rise to the property known as **inductive react-ance,** X_L, of the inductor. Since the inductive reactance restricts the current in the circuit, *it has the dimensions of resistance.* The current, I, in a pure inductive reactance X_L is calculated from the equation

$$\text{current}, I = \frac{V_L}{X_L} \text{ A} \qquad (11.3)$$

where V_L is the voltage across the inductor in volts, and X_L is the inductive reactance of the inductor in ohms. For example, if a resistanceless coil of 10 ohms inductive reactance is connected to a 240-V supply, then the current in the coil is

$$\text{current}, I = \frac{V_L}{X_L} = \frac{240}{10} = 24 \text{ A}$$

Turning now to the phase relationship between the voltage across the inductor and the current through it, your attention is directed to Figure 11.2. Since the supply voltage is connected to the resistanceless inductor, the voltage across the coil must be opposed by the back e.m.f. in the coil (this is clearly the case, since the coil has no resistance). You will recall from the earlier work on inductors that

'back' e.m.f. in an inductor = inductance × rate of change of current in the inductor

$$= L \frac{\Delta I}{\Delta t} \tag{11.4}$$

where $\frac{\Delta I}{\Delta t}$ is the 'rate of change of current'; that is the current changes by ΔI amperes in Δt seconds. To determine what the current waveshape itself looks like, consider the voltage waveforem in Figure 11.2(b). Eqn (11.4) relates the voltage across the inductor to the rate of change of current through it; we therefore need to deduce a method of working back from eqn (11.4) to the current waveshape. Since we may assume that the inductance of the coil has a constant value, eqn (11.4) reduces to

voltage across the inductor α rate of change (the 'slope') of the current waveform

where α means 'is proportional to', that is, doubling the voltage across the inductor doubles the rate of change of current in the inductor. Looking at the points A to E in Figure 11.2(b) we deduce the results in Table 11.1.

Bearing in mind the fact that the right-hand column in Table 11.1 is the *slope of the current waveform*, we deduce that the actual *shape of the waveform of current through L is a sinewave which lags by 90° behind the current waveform* (see Figure 11.2(c)).

Table 11.1 *Relationship between voltage and current waveforms in a pure inductor*

Point in figure 11.2(b)	Value of voltage	Slope of current waveform
A	zero	zero
B	positive (large)	positive (large)
C	zero	zero
D	negative (large)	negative (large)
E	zero	zero

The same solution can be found using **the calculus** which is included for the reader with a mathematical turn of mind. If the instantaneous supply voltage v_S is given by the expression

$v_S = V_{Sm} \sin \omega t$

where V_{Sm} is the maximum value of the supply voltage and ω is the angular frequency of the supply in rad/s, and the instantaneous value of the e.m.f., e, induced in the inductor is equal to v_S. Also $e = \frac{Ldi}{dt}$, where L is the inductance of the inductor in henrys and $\frac{di}{dt}$ is the rate of change of current through the coil in amperes per second and, since $e = v_S$, then

$$\frac{Ldi}{dt} = V_{Sm} \sin \omega t$$

$$di = \frac{V_{Sm}}{L} \sin \omega t \, dt$$

Integrating the equation with respect to time gives the following result for the instantaneous current, i, in the circuit

$$i = -\frac{V_{Sm}}{\omega L} \cos \omega t = \frac{V_{Sm}}{\omega L} \sin (\omega t - 90°)$$

$$= I_m \sin (\omega t - 90°) \tag{11.5}$$

where $(-\cos \omega t) = \sin (\omega t - 90°)$ and $I_m = \frac{V_{Sm}}{\omega L}$.

Equn (11.5) says that the current has a sinusoidal waveshape of maximum value I_m and **lags behind the voltage across the inductor by 90°**.

11.3 CALCULATION OF INDUCTIVE REACTANCE, X_L

The value of the **inductive reactance**, X_L, can be determined from the following relationship:

induced e.m.f. = inductance × rate of change of current in L

which can be rewritten in the form

average supply voltage = inductance L × average rate of change of current in the inductor

You will recall from Chapter 10 that the average value of a sinewave is $0.637V_m$, where V_m is the maximum value of the voltage. The average rate of change of current is calculated as follows. Since the maximum current I_m is reached after the first quarter cycle of the sinewave, the average rate of change of the current in this quarter cycle is $\frac{I_m}{T/4}$, where $\frac{T}{4}$ is one quarter of the periodic time of the cycle. Now, the periodic time of each cycle is $T = \frac{1}{f}$, where f is the supply frequency in Hz. That is,

$$\frac{1}{T/4} = \frac{1}{1/4f} = 4f$$

The equation for the average supply voltage therefore becomes

$$0.637V_m = L \times 4fI_m$$

or

$$\frac{V_m}{I_m} = \frac{4fL}{0.637} = 6.284fL = 2\pi fL = \omega L \tag{11.6}$$

where $\omega = 2\pi fL$ rad/sec. Now $X_L = \frac{V_L}{I}$, where V_L and I are the respective r.m.s. values of the voltage across L and the current in it. Also $V_L = 0.7071V_m$ and $I = 0.7071I_m$, then

$$XL = \frac{V_L}{I} = \frac{0.7071V_m}{0.7071\,I_m} = \frac{V_m}{I_m}$$

and since $\frac{V_m}{I_m} = \omega L$, then

inductive reactance, $X_L = \omega L = 2\pi fL$ Ω $\qquad\qquad$ (11.7)

Example
Calculate the current in a coil of 200 mH inductance if the supply voltage and frequency are 100 V and 500 Hz respectively.

Solution

$$L = 200 \text{ mH} = 0.2 \text{ H}; V_S = 100 \text{ V}; f = 500 \text{ Hz}$$

$$\text{inductive reactance}, X_L = 2\pi fL = 2\pi \times 500 \times 0.2$$

$$= 628.3 \ \Omega$$

From eqn (11.3)

$$\text{current}, I = \frac{V_S}{X_L} = \frac{100}{628.3} = 0.159 \text{ A (Ans.)}$$

11.4 X_L, I AND FREQUENCY

The equation of X_L is $2\pi fL$, which implies that *the reactance of a fixed inductance increases with frequency*, that is

$$\text{inductive reactance}, X_L \ \alpha \ \text{frequency}, f$$

This relationship is shown as a graph in Figure 11.3. At *zero frequency*, that is, at direct current, the value of X_L is zero. At an infinitely high frequency, the inductive reactance is infinitely large. This means that the inductor 'looks' like a short circuit to a d.c. supply, and at infinite frequency it looks like an open-circuit. This is illustrated by calculating the reactance of a 200 mH inductor as follows

fig 11.3 *inductive reactance, current and frequency*

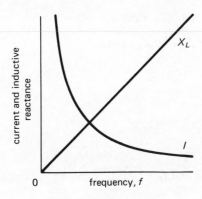

frequency	inductive reactance
zero	zero
500 Hz	628.3 Ω
5 kHz	6.283 kΩ
5 MHz	6.283 MΩ
5 GHz	6283 MΩ

The current, I, in a circuit containing only a pure inductor of reactance X_L is

$$I = \frac{V_S}{X_L}$$

Hence, as *the frequency increases, so the inductive reactance increases and the current reduces in value*. This is illustrated in Figure 11.3. At zero frequency $X_L = 0$ and $I = \frac{V_S}{0} = \infty$, and at infinite frequency $X_L = \infty$ and $I = \frac{V_S}{\infty} = 0$.

Example
An inductor has an inductive reactance of 10 Ω at a frequency of 100 Hz. Calculate the reactance of the inductor at a frequency of (i) 50 Hz, (ii) 400 Hz, (iii) 1500 Hz. If the supply voltage is 100 V r.m.s., determine also the current in the inductor for each frequency.

Solution
$$X_L = 10\ \Omega;\ f = 100\ Hz$$

If the inductor has a reactance of X_{L1} at frequency f_1 and X_{L2} at frequency f_2, then

$$X_{L1} = 2\pi f_1 L \text{ and } X_{L2} = 2\pi f_2 L$$

dividing the equation X_{L2} by X_{L1} gives

$$\frac{X_{L2}}{X_{L1}} = \frac{2\pi f_2 L}{2\pi f_1 L} = \frac{f_2}{f_1}$$

or

$$X_{L2} = X_{L1} \times \frac{f_2}{f_1}$$

If we let $X_{L1} = 10\ \Omega$ and $f_1 = 100$ Hz, then the reactances are calculated as follows

(i) $f_2 = 50$ Hz: $X_{L2} = X_{L1} \times \dfrac{f_2}{f_1} = 100 \times \dfrac{50}{100}$

$$= 50\ \Omega \text{ (Ans.)}$$

(ii) $f_2 = 400$ Hz: $X_{L2} = 100 \times \dfrac{400}{100} = 400\ \Omega$ (Ans.)

(iii) $f_2 = 1500$ Hz: $X_{L2} = 100 \times \dfrac{1500}{100} = 1500\ \Omega$ (Ans.)

The current in the inductor is calculated as follows.

(i) $I = \dfrac{V_S}{X_L} = \dfrac{100}{50} = 2$ A (Ans.)

(ii) $I = \dfrac{V_S}{X_L} = \dfrac{100}{400} = 0.25$ A (Ans.)

(iii) $I = \dfrac{V_S}{X_L} = \dfrac{100}{1500} = 0.0667$ A (Ans.)

11.5 CAPACITANCE IN AN a.c. CIRCUIT

We saw in Chapter 6 that when the voltage applied to a capacitor is increased, the capacitor draws a charging current (and when the voltage is reduced, the capacitor discharges). In an a.c. circuit the applied voltage is continually changing, so that the capacitor is either being charged or discharged on a more-or-less continuous basis.

Capacitive reactance

Since a capacitor in an a.c. circuit allows current to flow, it has the property of **capacitive reactance** which acts to restrict the magnitude of the current in the circuit. As with inductive reactance, *capacitive reactance has the dimensions of resistance*, that is, ohms. If V_C is the r.m.s. voltage across the capacitance and X_C is the capacitive reactance of the capacitor, then the equation for the r.m.s. current in the capacitor is

$$\text{current}, I = \frac{V_C \text{ (volts)}}{X_C \text{ (ohms)}} \tag{11.8}$$

For example, if a voltage of 20 V r.m.s. is applied to a capacitor and the current is 0.01 A, the capacitive reactance is

$$X_C = \frac{V_C}{I} = \frac{20}{0.01} = 2000 \ \Omega$$

Suppose that the capacitor current is changing in a sinusoidal manner as shown in Figure 11.4(b). It was shown in Chapter 6 that

capacitor charging current = capacitance, C × rate of change of voltage across
'the capacitor

$$= C \times \frac{\Delta V_C}{\Delta t}$$

where C is the capacitance of the capacitor in farads, and $\frac{\Delta V_C}{\Delta t}$ is the rate of change of the capacitor voltage in V/s. Since the capacitor is the only element in the circuit in Figure 11.4(a), the capacitor current is equal to the supply current. We can determine the waveshape of the capacitor voltage from the above relationship as follows. Since the capacitance, C, in the foregoing equation is constant, then

capacitor current α rate of change (slope) of the capacitor voltage

An inspection of the points A–E on the capacitor current wave in Figure 11.4(b) provides the data in Table 11.2. Bearing in mind that the right-hand column of the table is the *slope of the voltage waveform*, it can be deduced that the *shape of the waveform of the voltage across C is a sine-wave which lags 90° behind the current through C* (see Figure 11.4(c)). The corresponding phasor diagram is shown in Figure 11.4(d).

The following alternative solution for the phase angle of the capacitor voltage is obtained using the **calculus**. If the instantaneous current taken by the capacitor is given by $i = I_m \sin \omega t$, where I_m is the maximum value of the current through the capacitor, and ω is the angular frequency of the

fig 11.4 *pure capacitance in an a.c. circuit*

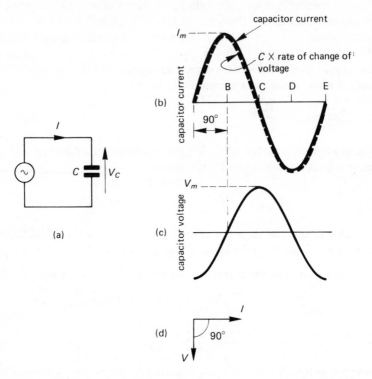

supply in rad/s, then the instantaneous voltage v_C across the capacitor is given by

$$i = C \times \frac{dv_C}{dt}$$

Table 11.2 *Relationship between current and voltage waveforms in a pure capacitor*

Point in figure 11.4(b)	Value of current	Slope of voltage waveform
A	zero	zero
B	positive (large)	positive (large)
C	zero	zero
D	negative (large)	negative (large)
E	zero	zero

where C is the capacitance of the capacitor in farads, and $\frac{dv_C}{dt}$ is the rate of change of the voltage across the capcitor in V/s. Since the current is sinusoidal, then

$$\frac{C\,dv_C}{dt} = I_m \sin \omega t$$

hence

$$dv_C = \frac{I_m}{C} \sin \omega t\,dt$$

Integrating the equation with respect to time results in the following equation for the instantaneous voltage, v_C, across the capacitor:

$$v_c = \frac{-I_m}{\omega C} \cos \omega t = \frac{I_m}{\omega C} \sin (\omega t - 90°)$$

$$= V_m \sin (\omega t - 90°) \qquad (11.9)$$

where $-\cos \omega t = \sin (\omega t - 90°)$ and $V_m = \frac{I_m}{\omega C}$. Eqn (11.9) says that the voltage across the capacitor is sinusoidal and has a maximum voltage V_m, and that *the voltage lags behind the current by 90°*.

11.6 CALCULATION OF CAPACITIVE REACTANCE, X_C

The value of the **capacitive reactance, X_C**, can be deduced from the following equation for the charging current:

capacitor current = capacitance × rate of change of voltage across C

which can be rewritten in the form

average current through capacitor = C × average rate of change of
voltage across capacitor

From Chapter 10, you will recall that the average value of a sinusoidal current is $0.637I_m$, where I_m is the maximum value of the current; this value will be used in the above equation.

The average rate of change of voltage across the capacitor is calculated as follows. Since the maximum voltage across the capacitor is V_m, and the time taken to reach this value is $\frac{T}{4}$ where T is the periodic time of the wave, the average rate of change of voltage during the first quarter cycle is $\frac{V_m}{T/4}$ or $V_m \times \frac{4}{T}$. Now the periodic time for each cycle is $T = \frac{1}{f}$, hence $\frac{T}{4} = \frac{1}{4f}$ therefore $\frac{V_m}{T/4} = \frac{V_m}{1/4f} = 4fV_m$. The above ewuation for the average current through the capacitor becomes

$$0.6371I_m = C \times 4fV_m$$

therefore

$$\frac{V_m}{I_m} = \frac{0.637}{4fC} = \frac{1}{6.28fC} = \frac{1}{2\pi fc} = \frac{1}{\omega C}$$

where $\omega = 2\pi f$. Now $X_C = \frac{V_C}{I}$, where V_C and I are the respective r.m.s. values of the voltage across and the current through the capacitor; also the r.m.s. voltage is given by $V_C = 0.7071V_m$, and the r.m.s. current by $I = 0.7071I_m$, hence

$$\text{capacitive reactance}, X_C = \frac{V}{I} = \frac{0.7071V_m}{0.7071I_m} = \frac{V_m}{I_m}$$

$$= \frac{1}{2\pi fC} = \frac{1}{\omega C} \ \Omega \qquad (11.10)$$

Example
A sinusoidal current of 0.1 A r.m.s. value flows through a capacitor connected to a 10-V, 1-kHz supply. Calculate the capacitance of the capacitor.

Solution

$$I = 0.1 \text{ A}; \ \ V_C = 10 \text{ V}; \ f = 1 \text{ kHz} = 1000 \text{ Hz}$$

From eqn (11.8), $I = \frac{V_C}{X_C}$, or

$$X_C = \frac{V_C}{I} = \frac{10}{0.1} = 100 \ \Omega$$

Now $X_C = \frac{1}{2\pi fC}$, hence

$$\text{capacitance}, C = \frac{1}{2\pi fX_C} = \frac{1}{(2\pi \times 1000 \times 100)}$$

$$= 1.592 \times 10^{-6} F \text{ or } 1.592 \ \mu F \text{ (Ans.)}$$

11.7 X_C, I AND FREQUENCY

Since the equation for X_C is $\frac{1}{2\pi fC}$, then for a fixed value of C *the capacitive reactance decreases with frequency*, or

$$\text{capacitive reactance}, X_C \ \alpha \ \frac{1}{\text{frequency}}$$

The relationship is shown in graphical form in Figure 11.5. At zero frequency, that is, direct current, the value of X_C is given by

$$X_C = \frac{1}{0} = \infty \ \Omega$$

fig 11.5 *capacitive reactance, current and frequency*

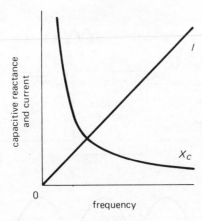

That is, the capacitor has infinite reactance to the flow of direct current.

As the frequency increases, X_C decreases and, at infinite frequency (which is beyond the scale of the graph in Figure 11.5), the capacitive reactance is

$$X_C = \frac{1}{\infty} = 0\ \Omega$$

That is the capacitor presents no opposition to the flow of very high frequency current. This can be summarised in the following.

At zero frequency (d.c.) the capacitor 'looks' like an open-circuit, and *at infinite frequency it 'looks' like a short-circuit*. For example a capacitor with a reactance of 100 Ω at a frequency of 1000 Hz has a reactance of 200 Ω at 500 Hz, and a reactance of 50 Ω at 2000 Hz.

The current in a circuit containing only a pure capacitance of reactance X_C is

$$I = \frac{V_C}{X_C} = \frac{V_C}{(1/2\pi f C)} = 2\pi C V_C \times f$$

That is, *as the frequency increases so the current through the capacitor increases*. At zero frequency the capacitive reactance is infinity, so that the current is

$$I = \frac{V_C}{\infty} = 0\ \text{A}$$

fig 11.6 *a full-wave rectifier and smoothing circuit with waveforms*

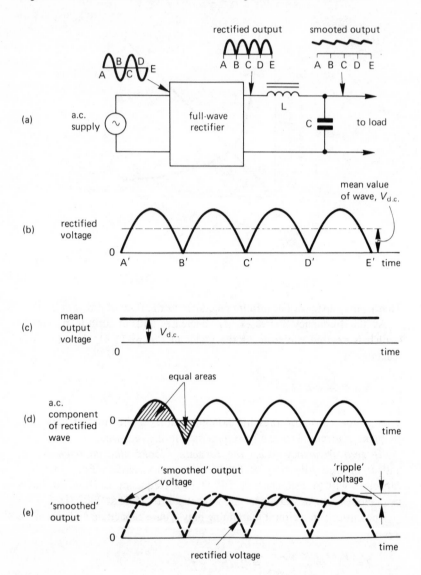

and at infinite frequency the capacitive reactance is zero, and the current is

$$I = \frac{V_C}{0} = \infty \text{ A}$$

11.8 SUMMARY OF CURRENT AND VOLTAGE RELATIONSHIPS

In an **inductor** we may say that EITHER

> the current through L **lags behind the voltage across it by 90°**,

OR

> the voltage across L **leads the current through it by 90°**,

In a **capacitor** we may say that EITHER

> the voltage across C **lags behind the current through it by 90°**,

OR

> the current through C **leads the voltage across it by 90°**.

These are neatly summarised by the mnemonic **CIVIL** as follows:

In C, I leads V

C I V I L

V leads I in L

11.9 APPLICATIONS OF INDUCTIVE AND CAPACITIVE REACTANCE

Whenever an inductor or capacitor is used in an a.c. circuit, its effect is felt in terms of its reactance. A popular application of both L and C are in a 'smoothing' circuit which acts to 'smooth out' the ripples in the output voltage from a rectifier circuit (a rectifier is a circuit which converts an alternating supply into a d.c. supply; the operation of rectifiers is fully described in Chapter 16 and need not concern us here).

The circuit in Figure 11.6 shows a **full-wave** rectifier which converts both positive and negative half-cycles of the alternating supply (hence the name full-wave) into a unidirectional or d.c. supply. During the positive half-cycles A–B and C–D of the a.c. wave, the rectifier produces a positive output voltage (shown in Figure 11.6(b) as A'–B' and C'–D', respectively). During the negative half cycles B–C and D–E of the a.c. wave, the rectifier once more produces a positive output voltage (shown as B'–C' and D'–E'). The resulting 'd.c.' output waveform is shown in Figure 11.6(b).

The latter waveform can be considered to consist of two part which can be added together to form the composite waveform in Figure 11.6(b). These parts are the 'steady' d.c. output voltage $V_{d.c.}$ (see Figure 11.6(c)), and an 'a.c.' component (Figure 11.6(d)). The composite rectified wave (Figure 11.6(b)) is applied to an *L-C* **ripple filter**; you will see that the output current from the rectifier must flow through L, and that C is connected between the output terminals.

It is important to note that L has little resistance and therefore does not impede the flow of d.c. through it.

The frequency of the 'ripple component' of the wave (which is the 'a.c.' component of the composite output) in Figure 11.6(d) has the following effect on the smoothing circuit:

1. the reactance of the inductor is fairly high (remember, inductive reactance increases with frequency);
2. the reactance of the capacitor is fairly low (remember, capacitive reactance reduces with increasing frequency).

The two reactance effects combine to reduce the 'ripple' voltage at the output terminals as follows. The high inductive reactance impedes the flow of ripple current through the reactor (this current can be regarded as 'a.c.' current) and the low reactance of the capacitor applies an 'a.c. short-circuit' to the output terminals. The former restricts the flow of ripple current to the d.c. load, and the latter by-passes the ripple current from the load. In this way, the *L-C* circuit **smooths out** the ripple at the output terminals of the circuit.

When an inductor is used in a smoothing circuit in the manner described above, it is known as a **choke** since it 'chokes' the flow of ripple current. The capacitor is sometimes described as a **reservoir capacitor** since it acts as a reservoir of energy during periods of time when the output voltage of the rectifier approaches zero; that is at A$'$, B$'$, C$'$, D$'$, etc. in Figure 11.6(b).

Another application of inductive and capacitive reactance is in the **tuning** of radios and televisions. A 'tuning' circuit consists of an inductor connected in parallel with a capacitor, one of the two components having a variable value. The tuning control has the effect of altering either the capacitance of the capacitor or the inductance of the inductor. When the circuit is 'tuned' to the desired frequency, the impedance to a.c. current flow is at its highest, so that for a given current flow the voltage across it is at its highest.

The impedance of the parallel circuit is lower to frequencies other than that to which it is tuned, so that it is less sensitive to these frequencies. In this way the tuning circuit rejects other frequencies than the one selected by the L and C of the circuit.

SELF-TEST QUESTIONS

1. The power consumed in an a.c. circuit containing a pure resistance is 200 W. If the supply voltage is 240 V, determine the current in the circuit and the resistance of the circuit.
2. What is meant by (i) inductive reactance, (ii) capacitive reactance? Circuit A contains an inductance of reactance 150 Ω and circuit B contains a capacitance of reactance 80 Ω. If the supply voltage is 100 V, calculate for each circuit (i) the current in the circuit and (ii) the phase angle of the circuit.
3. An inductance of 0.5 H is connected to a 100-V a.c. supply of frequency (i) 50 Hz, (ii) 60 Hz, (iii) 1 kHz. Calculate in each case the inductive reactance and the current in the circuit.
4. An a.c. supply of 10 V, 1 kHz is connected to a capacitor. If the current in the circuit is 0.628 A, calculate the capacitance of the capacitor.
5. The reactance of (i) an inductor, (ii) a capacitor is 100 Ω at a frequency of 850 Hz. Determine the reactance of each of them at a frequency of 125 Hz.

SUMMARY OF IMPORTANT FACTS

In a **pure resistive** circuit, the **current is in phase with the voltage across the resistor**. In a **pure inductive** circuit, the **current lags behind the voltage across L by 90°**. In a **pure capacitive** circuit the **current in the capacitor leads the voltage across it by 90°**. The mnemonic CIVIL is useful to remember these relationships (**in C, I leads V; V leads I in L**).

The **current** in a resistive circuit is given by $\frac{V_R}{R}$, and the **power consumed** is I^2R.

Inductive reactance, X_L, is given by the equation

$$X_L = 2\pi fL \quad (f \text{ in Hz}, L \text{ in henrys})$$

and the current in L is given by $I = \frac{V_L}{X_L}$. The **inductive reactance increases in proportion to the frequency**, that is $X_L \propto f$.

Capacitive reactance, X_C, is given by the equation

$$X_C = \frac{1}{2\pi fC} \quad (f \text{ in Hz}, C \text{ in farads})$$

and the current in C is given by $I = \frac{V_C}{X_C}$. The **capacitive reactance is inversely proportional to frequency**, that is $X_C \propto \frac{1}{f}$ (X_C reduces as f increases).

SINGLE-PHASE a.c.
CALCULATIONS

12.1 **A SERIES *R–L–C* CIRCUIT**

A single-phase circuit containing a resistor, an inductor and a capacitor is shown in Figure 12.1. You will recall that the phase relationship between the voltage and the current in a circuit element depends on the nature of the element, in other words, is it an R or an L or a C? This means that in an a.c. circuit you cannot simply add the *numerical* values of V_R, V_L and V_C together to get the value of the supply voltage V_S; the reason for this is that the *voltage phasors* representing V_R, V_L and V_C 'point' in different directions relative to the current on the phasor diagram. To account for the differing 'directions' of the phasors, you have to calculate V_S as the **phasor sum** of the three component voltages in Figure 12.1. That is

supply voltage, V_S = **phasor sum** of V_R, V_L and V_C

To illustrate how this is applied to the circuit in Figure 12.1, consider the case where the current, I, is 1.5 A, and the three voltages are

V_R = 150 V, V_L = 200 V, V_C = 100 V

fig 12.1 *an R–L–C series circuit*

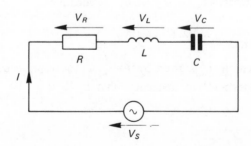

We shall consider in turn the phasor diagram for each element, after which we shall combine them to form· the phasor diagram for the complete circuit.

The phasor diagram for the *resistor* (Figure 12.2(a)) shows that the voltage V_R across the resistor is in phase with the current I. Phasor diagram (b) for the *inductor* shows that the voltage V_L across the inductor leads the current I (*remember* the mnemonic CIVIL). The phasor diagram for the *capacitor* (Figure 12.2(c)) shows the current I to lead the voltage V_C across the capacitor by 90°.

The **phasor diagram for the complete circuit** is obtained by combining the individual phasor diagrams. Since the current, I, is common to all three elements in the series circuit, it is drawn in the horizontal or **reference direction**. The combined phasor diagram is shown in Figure 12.2(d). First, V_L is shown leading I by 90°; next, V_R is added to V_L (remember, V_R is in phase with I, and therefore 'points' in the horizontal direction); next V_C is added to the sum of V_L and V_R to give the **phasor sum** of V_L, V_R

fig 12.2 *phasor diagrams for (a) R, (b) L and (c) C in Figure 12.1. The phasor diagram for the complete circuit is shown in (d)*

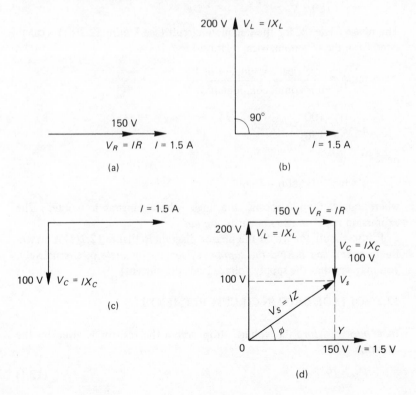

and V_C, which is equal to the supply voltage V_S. In engineering terms, the phasors V_L and V_C are in **quadrature** with I, meaning that they are at an angle of 90° to I or to the reference 'direction'. The supply voltage V_S has two components, namely its horizontal component and its quadrature component. These can be calculated from the phasor diagram as follows.

Since V_L and V_C do not have a horizontal component, then

horizontal component of $V_S = V_R = 150$ V

Since V_R does not have a quadrature component, the upward (quadrature) component of V_S is

quadrature component of $V_S = V_L - V_C$

$$= 200 - 100 = 100 \text{ V}$$

The triangle OV_SY in Figure 12.2(d) is a right-angled triangle, and from Pythagoras's theorem

$$V_S = \sqrt{[(\text{horizontal component})^2 + (\text{vertical component})^2]}$$

$$= \sqrt{[150^2 + 100^2]} = \sqrt{32500}$$

$$= 180.3 \text{ V}$$

The **phase angle**, ϕ, for the complete circuit (see Figure 12.2(d)) is calculated from the trigonometrical relationship

$$\tan \phi = \frac{\text{vertical component of } V_S}{\text{horizontal component of } V_S}$$

$$= \frac{100}{150} = 0.6666$$

hence

$$\phi = \tan^{-1} 0.6666 = 33.69°$$

where $\tan^{-1} 0.6666$ means 'the angle whose tangent is 0.6666'. The expression $\tan^{-1}$ is also known as 'arc tan'.

In the circuit for which the phasor diagram in Figure 12.2(d) is drawn, the *current I lags behind the supply voltage V_S by angle ϕ* (alternatively, you may say that the supply voltage leads the current).

12.2 VOLTAGE DROP IN CIRCUIT ELEMENTS

In a *pure resistor*, the voltage drop across the resistor is given by the equation

$$V_R = IR \tag{12.1}$$

where I is the r.m.s. value of the current. This is shown in the phasor diagram in Figure 12.2(a). For the values given in that case ($I = 1.5$ A, $V_R = 150$ V), the resistance is

$$R = \frac{V_R}{I} = \frac{150\,(\text{V})}{1.5\,(\text{A})} = 100\ \Omega$$

In a *pure inductive reactance*, the voltage drop across it is given by the equation

$$V_L = IX_L \tag{12.2}$$

and is illustrated in Figure 12.2(b). For the values given in the problem ($I = 1.5$ A, $V_L = 200$ V), the value of X_L is

$$X_L = \frac{V_L}{I} = \frac{200}{1.5} = 133.3\ \Omega$$

Finally, in a pure *capacitive reactance*, the voltage drop across the capacitor is

$$V_C = IX_C \tag{12.3}$$

This is shown in Figure 12.2(c); for the values given ($I = 1.5$ A, $V_C = 100$ V) the value of X_C is

$$X_C = \frac{V_C}{I} = \frac{100}{1.5} = 66.6\ \Omega$$

12.3 CIRCUIT IMPEDANCE, Z

The *total opposition* to the flow of alternating current of the circuit is known as the **impedance** of the circuit, symbol Z. That is, Z is the effective opposition to alternating current flow of all the components (R, L and C) in the circuit. The *magnitude* of the r.m.s. current flow, I, in an a.c. circuit connected to a sinusoidal voltage V_S is given by

$$I = \frac{V_S}{Z} \tag{12.4}$$

where Z is the impedance of the circuit.

Taking the values given in the example in Figure 12.2, the impedance of the circuit is

$$Z = \frac{V_S}{I} = \frac{180.3}{1.5} = 120.2\ \Omega$$

This value should be compared with the ohmic values of the elements in the circuit, namely

$$R = 100 \ \Omega$$

$$X_L = 133.3 \ \Omega$$

$$X_C = 66.6 \ \Omega$$

You should carefully note that the value of the impedance **is not equal to the sum of the ohmic values of the circuit components**; the impedance Z is calculated from R, X_L and X_C as follows.

Figure 12.3(a) shows the voltage components for the series circuit in Figure 12.1 (see also the phasor diagram in Figure 12.2(d)). The **voltage triangle** for the circuit comprises V_S, V_R and V_Q, where V_Q is the effective *quadrature voltage* in the circuit (that is $V_Q = V_L - V_C$). The value of each side of the voltage triangle is expressed in terms of

current $(I) \times$ ohmic value $(R, X$ or $Z)$

so that if you divide each side of the voltage triangle by I, you will be left with the **impedance triangle** of the circuit (see Figure 12.3(b)) showing the respective ohmic values of the circuit, namely $R, (X_L - X_C)$ and Z. You will also note that the **phase angle**, ϕ, of the circuit between voltage V_S and I is the same as the angle between Z and R in the impedance triangle. That is to say, you can calculate the phase angle of the circuit from a knowledge of the ohmic values in the circuit!

Since the impedance triangle is a right-angled triangle, the value of the impedance Z can be determined by Pythagoras's theorem as follows:

$$Z = \sqrt{[R^2 + (X_L - X_C)^2]} \tag{12.5}$$

fig 12.3 *(a) the voltage triangle for a series circuit and (b) the impedance triangle*

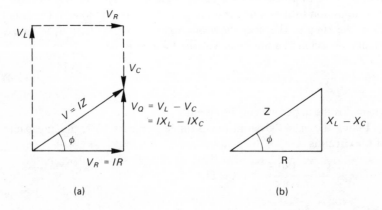

(a)   (b)

Using the values calculated for the series circuit in Figures 12.1 and 12.2
(R = 100 Ω, X_L = 133.3 Ω, X_C = 66.6 Ω), the impedance Z of the circuit
is calculated as follows

$$Z = \sqrt{[R^2 + (X_L - X_C)^2]}$$
$$= \sqrt{[100^2 + (133.3 - 66.6)^2]}$$
$$= \sqrt{[10\,000 + 4\,449.9]} = 120.2 \ \Omega$$

You will note that the above value is the same as that calculated from the
equation $Z = \frac{V_S}{I}$.

12.4 POWER IN AN a.c. CIRCUIT

It was shown in Chapter 10 that when *the current and voltage are 90° out
of phase with one another, the power consumed by the circuit is zero*. You
have already seen that the phase angle not only for a *pure inductor* but
also for a *pure capacitor* is 90°; that is

> the power consumed by a **pure inductor** and a **pure capacitor** is
> zero.

However, a *practical circuit* consists of a mixture of inductors, capacitors
and resistors, and the *power loss* in the circuit occurs **in the resistive parts**
of the circuit in the form of I^2R loss.

Moreover, since a *coil* is wound with wire which has resistance, *every
coil has some resistance*; consequently, there is some power loss in the *coil*
(there is, in fact, no power loss in the purely inductive part of the coil).
Also, as mentioned earlier, there is also some power loss in the coil if it has
an iron core attributed to eddy-current loss (due to current induced in the
iron core) and to hysteresis loss (due to the continued reversal of the magnetic domains as the alternating current reverses the direction of magnetisation of the magnetic field). The two latter types of power loss have the
effect of increasing the effective resistance of the coil. You must therefore
remember that a practical coil is far from being a perfect inductor.

Although a *practical capacitor* is a more-nearly 'perfect' element, it too
has some imperfections. Its dielectric is, in fact, not a perfect insulator,
but allows 'leakage' current to flow through it; that is, it has some 'leakage
resistance'.

A practical coil and a practical capacitor can be represented in the
forms shown in Figure 12.4. The coil is represented by inductor L in series
with resistor R which represents the combined effects of the resistance of
the coil, the eddy-current power-loss, and the hysteresis power-loss. A

fig 12.4 *the equivalent circuit of (a) a practical coil, (b) a practical capacitor*

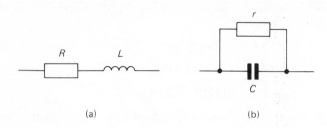

(a)   (b)

practical capacitor can be regarded as a pure capacitor C in parallel with resistor r, the resistance representing the 'leakage resistance'.

Consider now the series circuit in Figure 12.5 consisting of a resistor in series with a pure inductor (alternatively, you may think of the complete circuit as a single practical coil), connected to an a.c. supply. The circuit has an impedance Z of

$$Z = \sqrt{(R^2 + X_L^2)} = \sqrt{(10^2 + 15^2)} = 18.03 \ \Omega$$

The current I in the circuit is given by

$$I = \frac{V_S}{Z} = \frac{240}{18.03} = 13.31 \ \text{A}$$

Now, *the pure inductive element does not itself consume power*, so that the **power** consumed by the circuit is

$$\text{power}, P = I^2 R = 13.31^2 \times 10 = 1771.6 \ \text{W}$$

The **apparent power** consumed by the circuit is

$$\text{apparent power}, S = V_S I = 240 \times 13.31 = 3194.4 \ \text{VA}$$

and the **quadrature power** or VAr consumed is

$$\text{quadrature power}, Q = \sqrt{(S^2 - P^2)}$$
$$= \sqrt{(3194.4^2 - 1771.6^2)} = 2658.1 \ \text{VAr}$$

The above data allow us to calculate the **power factor** of the circuit as follows:

$$\text{power factor} = \frac{\text{'real' power}}{\text{apparent power}} = \frac{P}{S}$$
$$= \frac{1771.6}{3194.4} = 0.555$$

That is, 55.5 per cent of the volt–amperes consumed by the circuit are converted into watts.

fig 12.5 *a series a.c. circuit calculation*

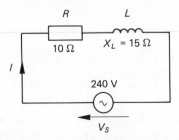

Now it was shown in Chapter 10 that the power factor of the circuit is equal to cos ϕ for the circuit, where ϕ is the phase angle between the supply voltage and the current drawn by the circuit. If we draw the phasor diagram for the circuit as shown in Figure 12.6, we see that

$$\cos \phi = \frac{V_R}{V_S} = \frac{IR}{IZ} = \frac{R}{Z}$$

that is for the circuit in Figure 12.5, the power factor is

$$\text{power factor} = \cos \phi = \frac{R}{Z} = \frac{10}{18.03} = 0.555$$

This shows that the power factor can be calculated either from a knowledge of the watts and volt–amperes, or from a knowledge of the resistance and reactance of the circuit elements.

12.5 PARALLEL a.c. CIRCUITS

A parallel circuit is one which has the same voltage in common with all its circuit elements. A typical parallel R–L–C circuit is shown in Figure 12.7. The phasor diagram for the complete circuit is deduced from the phasors

fig 12.6 *phasor diagram for an R–L series circuit*

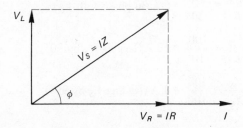

fig 12.7 *a parallel a.c. circuit*

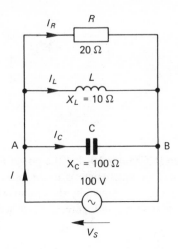

of the individual parallel branches. The calculation is kept as simple as possible by using 'pure' C and L elements in the bottom and centre branches of the circuit respectively.

The phasor diagram for the resistive, the inductive and the capacitive branches of the circuit are shown in diagrams (a), (b) and (c), respectively of Figure 12.8. In the case of a parallel circuit V_S is common to all branches, so that *the supply voltage is used as the reference phasor*, and is shown in the horizontal (reference) direction.

Applying Kirchhoff's first law to junction A (or to junction B) of Figure 12.7, and *bearing in mind that we are dealing with an a.c. circuit*, the supply current is given by the equation

$$\text{supply current}, I = \text{phasor sum} \ (I_R + I_L + I_C) \tag{12.6}$$

The complete phasor diagrams is obtained by adding the current phasors according to eqn (12.6) as shown in diagram (d) of Figure 12.8. The magnitude of the individual currents are calculated below:

$$I_R = \frac{V_S}{R} = \frac{100}{20} = 5 \text{ A (in phase with } V_S\text{)}$$

$$I_L = \frac{V_S}{L} = \frac{100}{10} = 10 \text{ A (lagging } V_S \text{ by } 90°\text{)}$$

$$I_C = \frac{V_S}{X_C} = \frac{100}{100} = 1 \text{ A (leading } V_S \text{ by } 90°\text{)}$$

fig 12.8 *phasor diagrams for (a) the resistive branch, (b) the inductive . branch, (c) the capacitive branch and (d) the complete circuit*

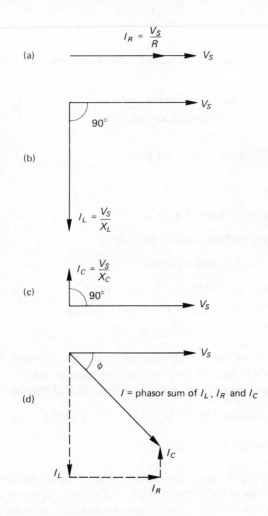

The *horizontal component, I_h*, of the total current is given by

horizontal component, $I_h = I_R$ p.245

The *vertically downwards component, I_v*, of the total current is

$$I_v = I_L - I_C = 10 - 1 = 9 \text{ A}$$

The magnitude of the total current, I, drawn from the supply is determined using Pythagoras's theorem as follows:

$$\text{total current, } I = \sqrt{(I_v^2 + I_v^2)} = \sqrt{(5^2 + 9^2)}$$

$$= 10.3 \text{ A}$$

You will see from Figure 12.8(d) that I *lags behind* V_S *by angle* ϕ, where

$$\tan \phi = \frac{\text{vertically downwards component of } I}{\text{horizontal component of } I}$$

$$= \frac{(I_L - I_C)}{I_R} = \frac{9}{5} = 1.8$$

hence

$$\text{phase angle, } \phi = \tan^{-1} 1.8 = 60.95°$$

The *power factor of the circuit* is given by

$$\text{power factor} = \cos \phi = \cos 60.95° = 0.4856$$

and the *power consumed by the complete circuit* is

$$\text{power, } P = V_S I \cos \phi = 100 \times 10.3 \times 0.4856$$

$$= 500 \text{ W}$$

Since the inductor and the capacitor are 'pure' elements they do not consume any power, and *all the power* is consumed in resistor R. To verify this fact we will calculate the power consumed in R as follows:

$$\text{power consumed in } R = I_R^2 R = 5^2 \times 20 = 500 \text{ W}$$

12.6 RESONANCE IN a.c. CIRCUITS

The word resonance means 'to reinforce a vibration or oscillation'. When resonance occurs in an electrical circuit, the circuit acts either to reinforce the current taken from the supply (**series resonance** or **acceptor resonance**) or to reinforce the current circulating between the branches of a parallel circuit (**parallel resonance** or **rejector resonance**).

The resonant condition occurs when the reactive elements (L and C) have values which cause the circuit to 'vibrate' electrically in sympathy with the electrical supply; this produces the reinforcement of either the voltage or current within the circuit.

The reason for resonance in an electrical circuit is, in fact, the interchange of the stored energy between the electromagnetic field of the inductor and the electrostatic field of the capacitor.

The electrical circuit 'resonates' at some frequency f_O Hz (or ω_O rad/s), and the current I_O is **in phase with the supply voltage** at the resonant frequency. That is, a circuit containing L and C appears at the resonant frequency as though it were a pure resistive circuit (see I_O in Figure 12.9). At any other frequency the current either leads the supply voltage (see I_1 in Figure 12.9). or it lags behind it (see I_2).

fig 12.9 *resonance in an a.c. circuit*

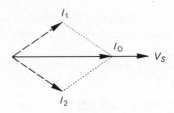

12.7 SERIES RESONANCE

From eqn (12.5) the impedance of a series circuit *at any frequency*, is given by

$$\text{impedance}, Z = \sqrt{[R^2 + (X_L - X_C)^2]}$$

If both sides of the equation are 'squared' it becomes

$$Z^2 = R^2 + (X_L - X_C)^2$$

At the resonant frequency (f_O or ω_O), the 'impedance' is purely resistive (it must be resistive since the current and voltage are in phase with one another!); that is, the 'reactive' part $(X_L - X_C)^2$ must be zero. This part of the equation can only be zero if $X_L = X_C$, that is, if

$$\omega_O L = \frac{1}{\omega_O C}$$

Hence the resonant frequency of the circuit is given by

$$\omega_O = \frac{1}{(LC)} \text{ rad/s} \tag{12.7}$$

where L is in henrys and C in farads. However, since $\omega_O = 2\pi f_O C$, the resonant frequency f_O is

$$f_O = \frac{1}{[2\pi\sqrt{(LC)}]} \tag{12.8}$$

For the circuit in Figure 12.10, $L = 0.1$ H and $C = 10$ μF, the resonant frequency is

$$\omega_O = \frac{1}{\sqrt{(LC)}} = \frac{1}{\sqrt{(0.1 \times [10 \times 10^{-6}])}} = 1000 \text{ rad/s}$$

and

$$f_O = \frac{\omega_O}{2\pi} = \frac{1000}{2\pi} = 159.2 \text{ Hz}$$

You will note that *the resistance of the circuit does not appear in any of the foregoing calculations since it does not affect the resonant frequency of the series circuit.*

As a matter of interest we will calculate the value of X_L and of X_C at resonance as follows:

$$X_L = \omega_O L = 1000 \times 0.1 = 100 \ \Omega$$

$$X_C = \frac{1}{\omega_O C} = \frac{1}{(1000 \times [10 \times 10^{-6}])} = 100 \ \Omega$$

fig 12.10 *series resonance*

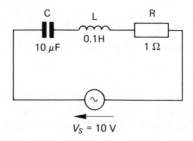

Resonance curves for a series circuit

At very low frequency, the capacitive reactance of the capacitor is very high (remember, $X_C = \frac{1}{2\pi f C}$), with the result that impedance of the series circuit is very high at low frequency. Consequently, the current flow in the circuit is small at low frequency (see Figure 12.11).

As the frequency of the supply increases, the capacitive reactance reduces in value, and with it the circuit impedance reduces in value. This results in an increase in current as the frequency increases in value (it being assumed for the moment that the supply voltage remains constant).

The current in a series circuit reaches its maximum value at the resonant frequency, when $Z = R$ and $I = \frac{V_S}{R}$.

fig 12.11 *frequency response of an R–L–C series circuit*

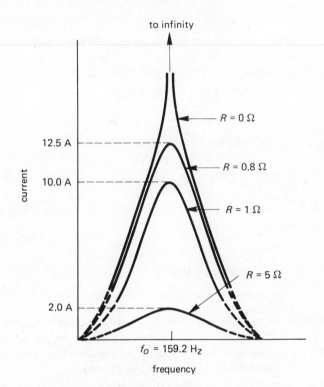

Beyond the resonant frequency there is a reduction in the circuit current (see Figure 12.11) for the following reason. As the frequency increases, the inductive reactance increases (remember, $X_L = 2\pi f L$) at a rate which is greater than the reduction in capacitive reactance. This results in a net increase in circuit impedance beyond the resonant frequency. Consequently, the current diminishes when the frequency passes through resonance.

At infinite frequency the inductive reactance X_L is infinitely large, so that the impedance of the circuit becomes infinity. That is, at infinite frequency the current diminishes to zero.

The curves in Figure 12.11 (known as the **frequency response curves**) show the variation in current in a series circuit for various values of frequency (the supply voltage being constant). You will see that the maximum r.m.s. value of current in the circuit is given by $I = \frac{V_S}{R}$; for a supply voltage of 10 V r.m.s., the resonant current in the circuit for a circuit resistance of 5 Ω is $\frac{10}{5} = 2$ A, for a resistance of 1 Ω is $\frac{10}{1} = 10$ A etc, and for a resistance of zero ohms is $\frac{10}{0} = \infty$ A.

12.8 *Q*-FACTOR OR QUALITY FACTOR OF A SERIES RESONANT CIRCUIT

We will illustrate the *Q*-**factor** or **voltage magnification factor** of a series circuit at resonance by means of the circuit in Figure 12.10. The impedance of the circuit *at its resonant frequency* of 1000 rad/s is

$$Z = \sqrt{[R^2 + (X_L - X_C)^2]} = \sqrt{R^2 + (100 - 100)^2]}$$

$$= R = 1 \ \Omega$$

The current drawn from the 10-V supply at resonance is

$$I = \frac{V_S}{Z} = \frac{10}{1} = 10 \ \text{A}$$

Since this current flows through *R*, *L* and *C*, the voltage across each of these elements is

$$V_R = IR = 10 \times 1 = 10 \ \text{V}$$

$$V_L = IX_L = 100 \times 10 = 1000 \ \text{V}$$

$$V_C = IX_C = 100 \times 10 = 1000 \ \text{V}$$

Although the supply voltage is only 10 V, the voltage across L and across C is 1000 V! That is, the circuit has 'magnified' the supply voltage 100 times! This is a measure of the 'quality' of the circuit at resonance; the value of this magnification is given the name **Q-factor**.

Let us determine how its value can be calculated. As you can see from the above calculations, the voltage across *R* is equal to the supply voltage, that is

$$V_S = V_R = IR$$

The voltage V_C across the capacitor is equal to IX_C, hence the *Q*-factor is

$$Q\text{-factor} = \frac{\text{voltage across } C \text{ at resonance}}{\text{voltage across } R \text{ at resonance}}$$

$$= \frac{IX_C}{IR} = \frac{X_C}{R} = \frac{1/\omega_O C}{R}$$

$$= \frac{1}{\omega_O CR} \qquad (12.9)$$

Similarly

$$Q\text{-factor} = \frac{\text{voltage across } L \text{ at resonance}}{\text{voltage across } R \text{ at resonance}}$$

$$= \frac{IX_L}{IR} = \frac{X_L}{R} = \frac{\omega_0 L}{R} \tag{12.10}$$

Referring to the circuit in Figure 12.10 and using eqn (12.9)

$$Q\text{-factor} = \frac{1}{\omega_O CR}$$

$$= \frac{1}{[1000 \times (10 \times 10^{-6}) \times 1]} = 100$$

It is left as an exercise for the reader to verify that eqn (12.10) gives the same result.

12.9 FEATURES OF SERIES RESONANCE

You can observe from the foregoing that **the total impedance of a series circuit is at its lowest value at the resonant frequency**, hence *the largest value of current flows in the circuit at resonance*. In the case of an *electronic circuit*, the resistance of the circuit is usually fairly high so that even at series resonance the current does not have a very high value.

On the other hand, in an electrical *power circuit* the resistance values are low (often only a fraction of an ohm), so that the current at resonance can have a dangerously high value. The high value of current produces not only damaging heating effects but also very high values of voltage across the inductors and capacitors in the circuit (remember, the voltage across these is Q times the supply voltage!) Both these effects can damage the circuit elements, and it is for this reason that **series resonance is avoided in power circuits**.

12.10 PARALLEL RESONANCE

Resonance occurs in the parallel circuit in Figure 12.12 when the total current I is in phase with the supply voltage V_S; it occurs at the resonant frequency ω_O rad/s or f_O Hz. The total current I is determined from the expression

total current, I = **phasor sum** of I_L and I_C

Now, the current in the branch containing the pure inductor is

$$I_L = \frac{V_S}{X_L}$$

and the current in the capacitive branch is

$$I_C = \frac{V_S}{X_C}$$

At the resonant frequency, ω_O, the reactive component of the total current is zero, that is $I_L = I_C$ or

$$\frac{V_S}{X_L} = \frac{V_S}{X_C}$$

that is

$$X_L = X_C$$

or

$$\omega_O L = \frac{1}{\omega_O C}$$

Cross-multiplying and taking the square root of both sides of the equation gives the resonant frequency ω_O as

$$\omega_O = \frac{1}{\sqrt{(LC)}} \text{ rad/s} \tag{12.11}$$

or

$$f_O = \frac{1}{2\pi\sqrt{(LC)}} \tag{12.12}$$

For the circuit in Figure 12.10

$$\omega_O = \frac{1}{\sqrt{(LC)}} = \frac{1}{\sqrt{(0.1 \times [10 \times 10^{-6}])}}$$

$$= 1000 \text{ rad/s}$$

or

$$f_O = \frac{\omega_O}{2\pi} = \frac{1000}{2\pi} = 159.2 \text{ Hz}$$

Comparing the equations for the resonant frequency for the series circuit (eqns (12.7) and (12.8)) and those for the simplified parallel circuit above, you will see that they are the same!. However, whilst it is true that the resistance of the series circuit does not affect the calculation for the resonant frequency of the series circuit, it does in fact affect the parallel circuit. This fact did not arise in the case of Figure 12.12 since it was assumed that the coil was a 'perfect' inductor. It can be shown that if the

fig 12.12 *parallel resonance*

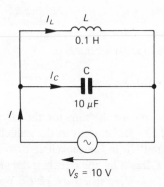

coil contains some resistance R, the equation for the resonant frequency of the parallel circuit is

$$\omega_O = \sqrt{\left(\frac{1}{LC} - \frac{R^2}{L^2}\right)} \text{ rad/s}$$

and

$$f_O = \frac{\omega_O}{2\pi} \text{ Hz}$$

If the resistance of the coil in Figure 12.12 is 1 Ω, the resonant frequency is calculated to be 999.95 rad/s or 159.15 Hz (compared with 1000 rad/s and 159.2 Hz, respectively), but if the resistance of the coil is 500 Ω, the resonant frequency is 866 rad/s or 137.8 Hz! Clearly, only if R is much less in value than X_L at the resonant frequency can its value be ignored.

12.11 CURRENT DRAWN BY A PARALLEL RESONANT CIRCUIT

The total current, I, drawn by a parallel resonant circuit at resonance is best determined by means of an example. Take the case of the circuit in Figure 12.12. The impedance of the inductive branch at the resonant frequency of 1000 rad/s is

$$X_L = \omega_O L = 1000 \times 0.1 = 100 \ \Omega$$

The impedance of the capacitive branch at the resonant frequency is

$$X_C = \frac{1}{\omega_O C} = \frac{1}{(1000 \times [10 \times 10^{-6}])} = 100 \ \Omega$$

The current, I_L, in the inductive branch is

$$I_L = \frac{V_S}{X_L} = \frac{10}{100} = 0.1 \text{ A}$$

and the current I_C in the capacitive branch is

$$I_C = \frac{V_S}{X_C} = \frac{10}{100} = 0.1 \text{ A}$$

Let us now look at the phasor diagram for the circuit at the resonant frequency in Figure 12.13. The current in the inductive branch lags 90° behind V_S, and the current I_C in the capacitive branch leads V_S by 90°. Now, the total current I drawn by the circuit is the *phasor sum* of I_L and I_C; *this is clearly seen to be zero in Figure 12.13.* We now have to solve the apparent dilemma of the circuit, namely that whilst a current flows in both L and C, no current flows in the main circuit!

fig 12.13 *current at resonance in a parallel circuit*

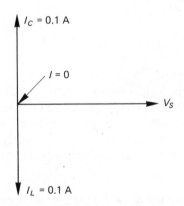

A reason for this is offered by the waveform diagram for the circuit in Figure 12.14. Under resonant conditions, the value of the current in each branch has the same value (see the calculation above) but, whilst I_C leads V_S by 90°, I_L lags behind V_S by 90°; the corresponding waveform diagrams are as illustrated in Figure 12.14. If we add I_L and I_C together at every instant of time, we see that **the total current I is zero**.

The value of the total current can also be arrived at by considering the energy flow in the circuit. During the period A–B in Figure 12.14, the capacitor is absorbing energy from the supply (I_C is positive!) and, at the same time, the inductor is returning energy to the supply (I_L is negative!). In effect, during the period A–B, the energy which is given up by L is

fig 12.14 *waveforms in a parallel resonant circuit*

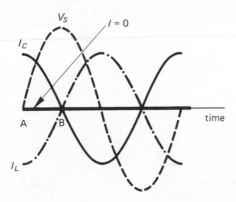

absorbed by C! Hence, during that period of time there is no need for the power supply to provide energy (current) to the circuit.

That is, **a parallel resonant circuit containing a pure inductor and a pure capacitor does not draw any current from the supply.**

Current drawn when the circuit contains resistance

In a practical circuit, the coil in the circuit contains some resistance (as does the wiring and the capacitor for that matter), which consumes electrical energy. This energy must be supplied by the power source in the form of a current; that is, a 'practical' parallel resonant circuit takes some current from the supply at the resonant frequency.

The resistance, R_D (known as the **dynamic resistance**), of a practical resonant circuit at resonance is given by the equation

$$R_D = \frac{L}{CR} \ \Omega$$

For example, if $L = 0.1$ H, $C = 10 \ \mu$F and $R = 1 \ \Omega$, then

$$R_D = \frac{L}{CR} = \frac{0.1}{([10 \times 10^{-6}] \times 1)} = 10\,000 \ \Omega$$

You may like to check that if $R = 10 \ \Omega$ then $R_D = 1000 \ \Omega$, and if $R = 100 \ \Omega$ then $R_D = 100 \ \Omega$. With a supply voltage of 10 V, the total current, I, drawn from the supply at the resonant frequency (calculated from the equation $I = \frac{V_S}{R_D}$) for various values of R is as listed in Table 12.1.

The current-frequency curves are shown in Figure 12.15. The interesting point to note about the curves is that the current drawn by the circuit at resonance *rises* as the *resistance of the inductive branch of the circuit*

Table 12.1 *Dynamic resistance and current for a parallel resonant circuit having resistance R ohms in its inductive branch*

Resistance, R ohms	Dynamic resistance, R_D ohms	Current at resonance, I amperes
0	∞	0
1	10 000	0.001
10	1 000	0.01
100	100	0.1

fig 12.15 *current-frequency graph for a parallel circuit having various resistance values*

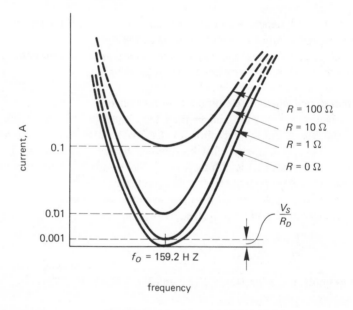

rises. This is explained by the fact that more energy is dissipated in the resistive branch when it has a higher resistance.

12.12 *Q*-FACTOR OF A PARALLEL RESONANT CIRCUIT

The quality factor or *Q-factor* of a parallel resonant circuit is equal to the *current magnification* that the circuit provides at resonance, as follows:

$$Q = \text{factor} = \frac{\text{current in } C \text{ (or in } L) \text{ at resonance}}{\text{current drawn from the supply at resonance}}$$

$$= \frac{I_C}{I} \text{ or } \frac{I_1}{I}$$

where I_1 is the current in the coil, that is, in the inductive branch. The Q-factors for the parallel circuit having the dynamic resistances listed in Table 12.1 are given in Table 12.2. You can also use the following equation to calculate the Q-factor for a practical parallel circuit containing some resistance in the inductive branch.

$$Q = \frac{\omega_O L}{R} \tag{12.13}$$

$$Q = \frac{1}{\omega_O C R} \tag{12.14}$$

Table 12.2 *Q-factor of a parallel resonant circuit*

$R(\Omega)$	$I(A)$	$IC(A)$	Q
0	0	0.1	∞
1	0.001	0.1	100
10	0.01	0.1	10
100	0.1	0.1	1

SELF-TEST QUESTIONS

1. A series $R-L-C$ circuit contains a 100-Ω resistor, a 0.5-H inductor and a 10-μF capacitor. If the supply frequency is 200 Hz, calculate the reactance of the inductor and of the capacitor, and the impedance of the circuit.

2. In question 1, if the supply voltage is 10 V r.m.s., calculate the current in the circuit and the voltage across each element in the circuit. Draw the phasor diagram of the circuit to scale. What is the power consumed and the power factor of the circuit?

3. The three elements in question 1 are reconnected in parallel with one another in the manner shown in Figure 12.7. If the supply voltage is 10 V, 200 Hz, calculate (i) the current in each branch of the circuit, (ii) the total current drawn by the circuit, (iii) the phase angle of the

total current with respect to the supply voltage and the circuit power factor, (iv) the VA, the power and the VAr consumed by the circuit. Draw the phasor diagram of the circuit to scale.

4. What is meant by resonance in an a.c. circuit? Explain under what conditions (i) series resonance, (ii) parallel resonance occurs.
5. A series circuit containing a resistor of 10 Ω resistance, an inductor of 0.05 H inductance and a capacitor resonates at a frequency of 1 kHz. Calculate the capacitance of the capacitor. If the supply voltage is 15 V, determine the current in the circuit at resonance.
6. Determine the Q-factor of the circuit in question 5.

SUMMARY OF IMPORTANT FACTS

The **impedance** of a circuit is its **total opposition** to flow of current.

The **voltage triangle** of a series circuit shows the voltage across the resistive elements (IR), and across the reactive elements ($IX = IX_L - IX_C$), and across the complete circuit (IZ). The **phase angle** of the circuit is the angle between the supply voltage and the current drawn from the supply, and can be determined from the angle in the voltage triangle. The **impedance triangle** for a circuit is obtained by dividing each side of the voltage triangle by the current, I, and shows the total resistance, the effective reactance, and the impedance of the circuit.

The **volt-amperes** (VA) or **apparent power** consumed by an a.c. circuit is equal to the product of the supply voltage and the supply current. The **power consumed** is either equal to the sum of the I^2R products in the circuit (the power loss in all or the resistors), or is equal to $V_S I \cos \phi$, where V_S is the supply voltage, I is the current drawn from the supply and ϕ is the phase angle between V_S and I.

In a **series circuit**, the **phasor sum** of the voltages across the circuit elements is equal to the supply voltage. In a **parallel circuit**, the **phasor sum** of the current in each of the branches is equal to the supply current.

Resonance occurs in an R–L–C circuit when the current drawn from the supply is in phase with the supply voltage. A *series resonant circuit* is known as an *acceptor circuit*; a *parallel resonant circuit* is known as a *rejector circuit*.

The **impedance** of a *series resonant circuit* is equal to the resistance of the circuit. The **impedance** of a *parallel resonant circuit* is known as the **dynamic impedance** of the circuit and is equal to $\frac{L}{CR}$ ohms, where R is the resistance in the inductive branch of the circuit; the dynamic impedance of a parallel circuit usually has a very high value.

The **Q-factor** of a resonant circuit is a measure of the *quality* of the circuit in terms of magnifying either the voltage (series resonance) or the current (parallel resonance).

POLY-PHASE a.c. CIRCUITS

13.1 FEATURES OF A POLY-PHASE SUPPLY

As its name implies, a poly-phase power supply or multi-phase supply provides the user with several power supply 'phases'. The way in which these 'phases' are generated is described in sections 13.2 and 13.3, and we concentrate here on the advantages of the use of a poly-phase supply which are:

1. For a given amount of power transmitted to the user, the volume of conductor material needed in the supply cable is less than in a single-phase system to supply the same amount of power. A poly-phase transmission system is therefore more economical than a single-phase supply system.
2. Poly-phase motors and other electrical equipment are generally smaller and simpler than single-phase motors and equipment. For industry, poly-phase equipment is cheaper and easier to maintain.

A poly-phase supply system may have two, three, four, six, twelve or even twenty-four phases, with the **three-phase system** being the most popular. The National Grid distribution network is a three-phase system. An introduction to electrical power distribution systems was given in Chapter 8, where it was shown that power is distributed to industry using a three-phase system, a single-phase system being used for domestic power distribution.

13.2 A SIMPLE TWO-PHASE GENERATOR

Consider the simple single-phase generator or **alternator** in Figure 13.1(a). If line N1 is earthed, the waveform of the alternating voltage on line L is shown dotted in Figure 13.1(c).

fig 13.1 *a simple two-phase generator*

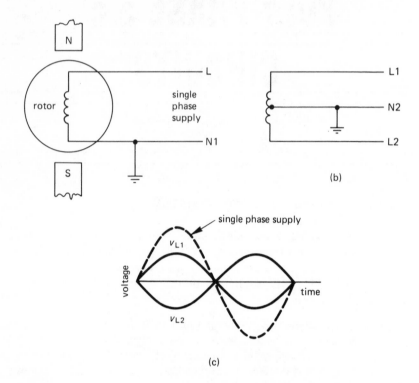

(b)

(c)

Suppose now that we disconnect the earth from line N1 and move it to the centre-point of the alternator winding (shown as N2 in diagram (b)). In effect, the two 'ends' of the winding are 'live' with respect to point N2, but they each have one-half the r.m.s. voltage on them when compared with the single-phase case (Figure 13.1(a)). Moreover, when line L1 is positive with respect to point N2, then L2 is negative with respect to it. The corresponding waveform diagrams for the voltages on lines L1 and L2 are shown in Figure 13.1(c).

Clearly, **by earthing the centre point of the alternator winding, we have produced two sets of 'phase' voltages which can be used independently and have (in this case) a phase angle difference of 180° between them.** This type of power supply is used in many bi-phase rectifier circuits (see Chatper 16).

13.3 A THREE-PHASE GENERATOR

A three-phase supply can be thought of as being generated by three windings on the rotor of an alternator connected as shown in Figure

13.2(a) [you should note that this is only one of several possible methods of connection]. The voltage induced in each winding or **phase** is known as the **phase voltage**, and the three voltages are described as follows:

V_{RN} is the **red phase** voltage

V_{YN} is the **yellow phase** voltage

V_{BN} is the **blue phase** voltage

fig 13.2 *a three-phase generator*

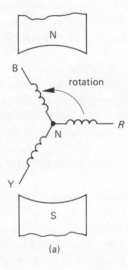

(a)

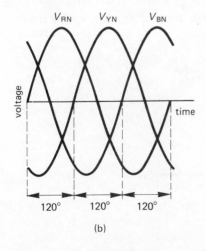

(b)

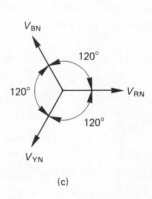

(c)

The *N*-point in Figure 13.2(a) is known as the 'neutral' point since it is often connected to earth, that is, to a 'neutral' potential.

Since each winding on the alternator rotor is displaced from each of the other windings by 120°, the phase angle between each of the generated waveforms (Figure 13.2(b)) is 120°; the corresponding phasor diagrams are shown in Figure 13.2(c). With the conventional direction of rotation in Figure 13.2(a), the voltages become positive in the sequence red, yellow, blue; the **phase sequence R, Y, B** is known in electrical engineering as the **positive phase sequence**.

The phase sequence of a poly-phase electrical supply can be deduced as follows. Imagine that you are standing on the right-hand side of the alternator rotor in Figure 13.2(a) and looking towards point N. With the direction of rotation shown for the rotor, the windings pass you in the sequence R, Y, B; this means that the voltages become positive in the sequence R, Y, B.

If the direction of the alternator rotor is reversed, the phase sequence is R, B, Y; this means that the phase voltages become positive in the sequence R, B, Y. The latter sequence is known as **negative phase sequence**. You would find it an interesting exercise to draw the waveform diagrams for an alternator having the phase sequence R, B, Y.

13.4 'BALANCED' AND 'UNBALANCED' THREE-PHASE SYSTEMS

A **balanced three-phase supply system** is one which has *equal values of phase voltage which are displaced from one another by 120°*. The phasor diagram in Figure 13.2(c) is for a balanced three-phase supply.

An **unbalanced three-phase supply** either has *unequal phase voltages, or has three voltages which are displaced from one another by an angle which is not 120°, or both*. The phasor diagram in Figure 13.3 is one for an unbalanced three-phase system; certain types of electrical loading and fault conditions produce unbalanced operation.

fig 13.3 *an unbalanced three-phase supply*

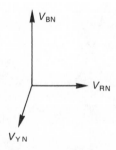

13.5 THE THREE-PHASE, FOUR-WIRE STAR CONNECTION

The **star connection**, as its name implies, means that the three windings of
the alternator (or of the connected load) have a common point or **star
point**. In Figure 13.4 both the generator and the load are 'star' connected
but, for clarity, we have described the 'star' point of the generator as the
neutral point (N) since it is usually connected to earth. The star point of
the load is given the symbol S.

Figure 13.4 shows a **three-phase, four-wire** supply system since, as its
name implies, four wires are used to connect the alternator to the load.
The fourth wire is the **neutral wire**, which is used to carry current between
the S-point of the load and the N-point of the generator. The three-phase,
four-wire supply system is used to supply electricity to houses within a
town or village. One group of houses would be supplied by power from the
'red' phase (strictly speaking, the power is supplied from the red–to–
neutral (R–N) phase), a second group of houses would be supplied from
the 'yellow' phase, and the remaining houses from the 'blue' phase. In the
United Kingdom the 'phase voltage' supplied to houses is 240 V, 50 Hz; in
the United States of America it is 110 V, 60 Hz.

As in any walk of life, anyone in a house is an individual and each has
his own demand on the electricity system. It follows that each group of
houses requires a different current at a different power factor (the latter
depending on the type of load that is switched on) than the other groups.
For example, the group of houses connected to the R-phase may need a
current of 100 A, the group connected to the Y-phase may need 20 A, and
the group connected to the B-phase may need 70 A. This load is *unbalanced*,
and it is the *phasor sum* of these currents which returns to the N-point of

fig 13.4 *the three-phase, four-wire star connection*

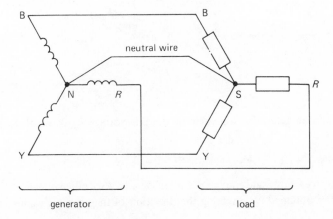

the generator along the neutral wire. This is one reason why the domestic power supply system needs the neutral wire or 'fourth' wire between the load and the supply.

Line voltage of a star connected system

When specifying a three-phase supply voltage, it is usual to quote the voltage between a pair of lines in the supply system; the **line-to-line** voltage is shortened to **line voltage**. The line voltages are specified as follows (see also Figure 13.5):

V_{RY} = voltage of line R relative to line Y

$\quad = V_{RN} - V_{YN}$

V_{BR} = voltage of line B relative to line R

$\quad = V_{BN} - V_{RN}$

V_{YB} = voltage of line Y relative to line B

$\quad = V_{YN} - V_{BN}$

fig 13.5 *phase and line voltage and currents in a star-connected system*

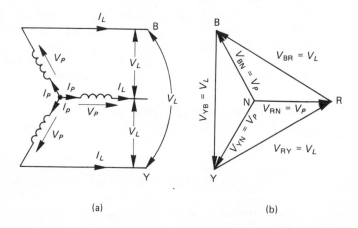

(a)   (b)

We will now determine the value of the line voltage V_{RY}, which is given by

$$V_{RY} = V_{RN} - V_{YN} = V_{RN} + (-V_{YN})$$

That is, we simply add the 'negative' of V_{YN} to V_{RN}; the voltage $(-V_{YN})$ is simply obtained by reversing the direction of the phasor V_{YN} as shown in Figure 13.6. By drawing the diagram accurately to scale you will find that the value of V_{RY} is:

$$V_{RY} = 1.732V_{RN} = \sqrt{3} \times V_{RN}$$

fig 13.6 *determining the line voltage V_{RY}*

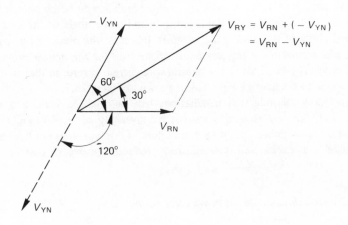

In general, we may say that

line voltage = $\sqrt{3}$ × phase voltage

or

$$V_L = \sqrt{3} \times V_P \tag{13.1}$$

For a phase voltage of 240 V (which is the nominal UK single-phase domestic voltage), the line voltage is

$$V_L = \sqrt{3} \times 240 = 415.7 \text{ V}$$

As mentioned ealier, three-phase supply systems are specified in terms of their line voltage, and we would describe the above supply as a 415.7 V, three-phase supply. The UK 400-kV grid system therefore operates with a line voltage of 400 kV and a line-to-neutral voltage of $\frac{400}{\sqrt{3}} = 230.9$ kV.

Line current of a star connected system

Since the current in one phase of the load flows in the supply line which is connected to it (see Figure 13.5), **the phase current, I_P, is equal to the line current, I_L**. That is

phase current = line current

or

$$I_P = I_L \qquad (13.2)$$

13.6 THREE-PHASE, THREE-WIRE STAR CONNECTED SYSTEM

If the load connected to a three-phase system is *balanced*, that is the load has *the same impedance in each of its three branches and each branch has the same phase angle*, then the magnitude and phase angle of the current in each phase of the load is the same. In this case, *the phasor sum of the three phase currents at the star point of the load (or the neutral point of the supply) is zero*. That is, **the neutral wire current is zero**, so that **we can dispense with the neutral wire!** This is shown in Figure 13.7.

Practically all industrial installations have a balanced load, and the three-phase, three-wire supply system is universally adopted in industry. Reducing the number of supply conductors in this way does not have any technical drawbacks, and it significantly reduces the overall cost of the supply system.

fig **13.7** *the three-phase, three-wire star connection*

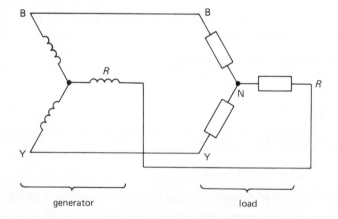

generator   load

13.7 THREE-PHASE DELTA OR MESH CONNECTION

If the three-phase generator described in section 13.3 has three separate windings which can be connected together in any way, and if we connect the 'end' of one winding to the 'start' of the next winding, we get the **delta connection** or **mesh connection** shown in Figure 13.8.

Whilst it may appear in this case that we are 'short-circuiting' the windings together, you must remember that the line voltages are at dif-

fig 13.8 *the delta or mesh connection of a three-phase generator and load*

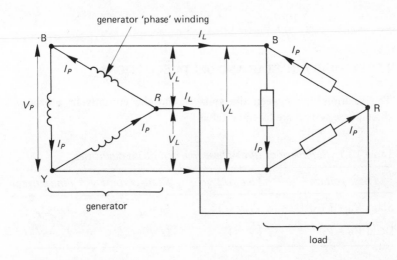

ferent phase angles to one another (120° in the case of a three-phase system), so that the voltage between the 'end' of the final voltage in the mesh and the 'beginning' of the first voltage is zero! Hence, when the mesh is connected as shown, no current circulates around the close mesh.

You will see that each winding of the generator (the 'phase' winding in which the phase voltage V_P is induced) is connected to two of the supply lines. That is, in a mesh or delta system

phase voltage = line voltage

or

$$V_P = V_L \qquad (13.3)$$

That is, if the *line voltage* of a distribution system is 6.6 kV, it can be supplied from a delta-connected generator with a 'phase' voltage of 6.6 kV (if the generator was star-connected, the phase voltage of the generator would be $\frac{6.6}{\sqrt{3}}$ = 3.81 kV).

For a *balanced delta-connected load*, the relationship between the *phase current* I_P in the phase winding (or in one phase of the load), and the *line current* I_L in the supply line is

line current = $\sqrt{3}$ × phase current

or

$$I_L = \sqrt{3} \times I_P$$

For example, if the line current supplied to a delta-connected machine is 100 A, the phase current is

$$I_P = \frac{I_L}{3} = \frac{100}{\sqrt{3}} = 57.74 \text{ A}$$

13.8 SUMMARY OF STAR AND DELTA EQUATIONS

The relationship between the voltages and the currents in a balanced three-phase system are listed in Table 13.1.

Table 13.1 *Summary of three-phase line and phase quantities*

	Phase voltage	Line voltage	Phase current	Line current
Star	$V_P = V_L\sqrt{3}$	$V_L = \sqrt{3}V_P$	$I_P = I_L$	$I_L = I_P$
Delta	$V_P = V_L$	$V_L = V_P$	$I_P = I_L/\sqrt{3}$	$I_L = \sqrt{3}I_P$

13.9 POWER CONSUMED BY A THREE-PHASE LOAD

In *any three-phase load, either balanced or unbalanced*, **the total power consumed is equal to the sum of the power consumed in each of the three phases.**

Example
Calculate the total power consumed in a three-phase system having a phase voltage of 1000 V and the following phase current and power factor:

        Red phase   10 A at 0.8 power factor

        Yellow phase   20 A at 0.2 power factor

        Blue phase   15 A at 0.6 power factor

Solution
The power consumed by each phase is given by the expression $V_P I_P \cos \phi_P$, where V_P is the phase voltage, I_P is the phase current, and $\cos \phi_P$ is the phase power factor. The power consumed in each phase is calculated as follows:

        power in R phase = $1000 \times 10 \times 0.8 = 8000$ W

        power in Y phase = $1000 \times 20 \times 0.2 = 4000$ W

        power in B phase = $1000 \times 15 \times 0.6 = 9000$ W

and

$$\text{total power consumed} = 8000 + 4000 + 9000$$

$$= 21\,000 \text{ W (Ans.)}$$

13.10 VA, POWER AND VAr CONSUMED BY A BALANCED THREE-PHASE LOAD

In the case of a *balanced load*, the current and the power factor associated with each phase of the load is the same. This allows us to deduce some general equations as follows. In the following

V_P = phase voltage

V_L = line voltage

I_P = phase current

I_L = line current

$\cos\phi$ = power factor of the load

The reader may find it helpful to consult Table 13.1 when reading the following.

VA consumed by a balanced load
The volt-amperes consumed *by one phase* of the load is $V_P I_P$, and the *total VA* consumed by the load is $3V_P I_P$.

Star connected system

$$\text{total VA}, S = 3V_P I_P = 3 \times \frac{V_L}{\sqrt{3}} \times I_L = \sqrt{3}V_L I_L \text{ VA}$$

Delta connected system

$$\text{total VA}, S = 3V_P I_P = 3 \times V_L \times \frac{I_L}{\sqrt{3}} = \sqrt{3}V_L I_L \text{ VA}$$

Power consumed by a balanced load
The power consumed by *one phase* is $V_P I_P \cos\phi$, and the total power consumed is $3V_P I_P \cos\phi$.

Star connected load

$$\text{total power}, P = 3V_P I_P \cos\phi = 3 \times \frac{V_L}{\sqrt{3}} \times I_L \cos\phi$$

$$= \sqrt{3}V_L I_L \cos \phi = S \cos \phi \text{ W}$$

Delta connected system

$$\text{total power}, P = 3V_P I_P \cos \phi = 3 \times V_L \times \frac{I_L}{\sqrt{3}} \cos \phi$$
$$= \sqrt{3}V_L I_L \cos \phi = S \cos \phi \text{ W}$$

Volt–amperes reactive consumed by a balanced load
The VAr consumed by *one phase* is $V_P I_P \sin \phi$, and the total VAr consumed is $3V_P I_P \sin \phi$.

Star connected load

$$\text{total VAr}, Q = 3V_P I_P \sin \phi = 3 \times \frac{V_L}{\sqrt{3}} \times I_L \sin \phi$$
$$= \sqrt{3}V_L I_L \sin \phi = S \sin \phi \text{ Var}$$

Delta connected load

$$\text{total VAr}, Q = 3V_P I_P \sin \phi = 3 \times V_L \times \frac{I_L}{\sqrt{3}} \sin \phi$$
$$= \sqrt{3}V_L I_L \sin \phi = S \sin \phi \text{ VAr}$$

Summary
In any three-phase balanced load

$$\text{total VA consumed}, S = \sqrt{3}V_L I_L \text{ VA} \tag{13.5}$$

$$\text{total power consumed}, P = \sqrt{3}V_L I_L \cos \phi \text{ W} \tag{13.6}$$

$$\text{total VAr consumed}, Q = \sqrt{3}V_L I_L \sin \phi \text{ VAr} \tag{13.7}$$

Example
A three-phase motor has a mechanical output power of 10 kW and an efficiency of 85 per cent. If the motor is supplied at a line voltage of 416 V and the power factor of the motor is 0.8 lagging, calculate the line current drawn by the motor. Calculate the phase current and voltage of the motor if it is (a) star connected, (b) delta connected.

Solution
Output power = 10 kW = 10 000 W; efficiency = 0.85; V_L = 416 V; $\cos \phi$ = 0.8.

The power input to the motor is

$$\text{power input} = \frac{\text{power output}}{\text{efficiency}} = \frac{10\,000}{0.85}$$

$$= 11\,765 \text{ W}$$

An electrical motor is a balanced load, so that we may use eqn (13.6) as follows

electrical input power to the motor $= \sqrt{3}V_L I_L \cos \phi$

hence

$$\text{line current, } I_L = \frac{P}{(\sqrt{3}V_L \cos \phi)}$$

$$= \frac{11\,765}{(\sqrt{3} \times 416 \times 0.8)}$$

$$= 20.41 \text{ A (Ans.)}$$

(a) For a *star connected* motor

$$I_P = I_L = 20.41 \text{ A (Ans.)}$$

$$V_P = \frac{V_L}{\sqrt{3}} = \frac{416}{\sqrt{3}} = 240 \text{ V (Ans.)}$$

(b) For a *delta connected* motor

$$I_P = \frac{I_L}{\sqrt{3}} = \frac{20.41}{\sqrt{3}} = 11.78 \text{ A (Ans.)}$$

$$V_P = V_L = 416 \text{ V (Ans.)}$$

SELF-TEST QUESTIONS

1. Explain why a poly-phase power distribution is used for the National Grid distribution system.
2. Describe the operation of a three-phase generator.
3. What is meant by (i) a balanced supply voltage and (ii) an unbalanced supply.
4. A three-phase supply has a line voltage of 33 kV; calculate the phase voltage if it is star connected. What is the line voltage of a star connected system which has a phase voltage of 100 V?
5. Under what circumstances are the following used? (i) A three-phase, four-wire system, (ii) a three-phase, three-wire system.

6. A factory draws a balanced three-phase load of 100 kW at a power factor of 0.8, the line voltage being 3.3 kV. Calculate the line current and the phase current if the load is (i) star connected, (ii) delta connected.

7. For the factory in question 6, calculate the phase angle of the load, the apparent power consumed, and the volt–amperes reactive consumed.

SUMMARY OF IMPORTANT FACTS

A **poly-phase supply** provides the user with several (usually three) power supply 'phases', the phase voltages being at a fixed angle to one another (which is 120° in a three-phase system). A poly-phase system can transmit more power for a given amount of conductor material than can an otherwise equivalent single-phase system. Poly-phase motors are generally smaller and more efficient than equivalent single-phase motors; polyphase switchgear and control gear is also smaller and cheaper than equivalent rated single-phase equipment.

A **balanced** or **symmetrical** three-phase supply has three equal phase voltages which are displaced from one another by 120°. The National Grid distribution network and local area distribution systems employ a three-phase distribution system.

Factories and other installations with three-phase **balanced loads** use a **three-phase, three-wire** distribution system. Housing estates and other consumers with an **unbalanced load** use a **three-phase, four-wire** distribution system; the fourth or **neutral wire** is used to return the out-of-balance current in the phases to the generator.

Generators and loads can be connected in one of several ways including **delta (mesh)** or **star**; star connected systems can use a three-phase, four-wire supply. In **star**, the line and phase currents are equal to one another; in **delta** or **mesh**, the line voltage is equal to the phase voltage.

The **power consumed** by a three-phase load (either balanced or unbalanced) is equal to the sum of the power consumed in each of the three loads. The *power consumed in a balanced load* is equal to $\sqrt{3} V_L I_L \cos \phi$.

THE TRANSFORMER

14.1 MUTUAL INDUCTANCE

In Chapter 7 (Electromagnetism) it was shown that

1. when a current flows in a coil, a magnetic flux is established;
2. when a magnetic flux cuts a coil of wire, an e.m.f. is induced in the coil.

The above effects are involved in the process of **mutual induction**, illustrated in Figure 14.1, in which a changing alternating current in one coil of wire (the **primary winding**) induces an e.m.f. in the second coil (the **secondary winding**). The general principle of operation is described below.

When a current flows in inductor L_1, it produces magnetic flux Φ. For practical reasons, not all this flux manages to link with the secondary winding; let us say that a proportion $k\Phi$ of this flux, where k is a number in the range zero to unity, links with the secondary winding. The flux which is common to the two windings is known as the **mutual flux**. This flux acts to induce mutually an e.m.f. V_2 in inductor L_2. The magnitude of V_2 is given by the formula

$$V_2 = M \frac{\Delta I_1}{\Delta t} \tag{14.1}$$

where M is the **mutual inductance** in henrys between the two coils, and $\frac{\Delta I_1}{\Delta t}$ is the rate of change of current I_1 with respect to time (in A/s). For example, if the primary winding current changes at the rate of 80 A/s (corresponding approximately to a sinewave of peak current value 1 A at a frequency of 50 Hz), and if the mutual inductance is 100 mH, the induced e.m.f. is

$$V_2 = M \Delta I_1/\Delta t$$

$$= (100 \times 10^{-3}) \times 80 = 8 \text{ V}$$

fig 14.1 *mutual inductance: the basis of the transformer*

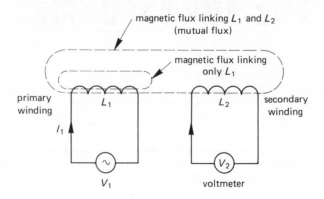

This is the principle of the **electrical transformer** in which the primary winding carries alternating current from the power supply, which is 'transformed' to a different voltage and current level by the mechanics of mutual inductance.

14.2 COUPLING COEFFICIENT

We saw in Figure 14.1 that not all the magnetic flux produced by the primary winding links with the secondary winding. The amount of magnetic flux linking with the secondary winding is $k\Phi$, where k is known as the **magnetic coupling coefficient** and is defined by the equation

$$k = \frac{\text{amount of flux linking the secondary winding}}{\text{amount of flux produced by the primary winding}}$$

$$= \frac{k\Phi}{\Phi}$$

The coupling coefficient has a value of zero when none of the flux produced by the primary winding links with the secondary winding, and has a value of unity when all the flux links with the secondary winding; the coupling coefficient is a dimensionless number.

Depending on the application, k may have a low value (a low value of k is used in certain radio and television 'tuning' circuits when k may be as low as 0.05) or it may have a high value (as occurs in electrical power transformers when k may be as high as 0.95, where we need a high electrical efficiency).

The mutual inductance, M henrys, between two coils of self inductance L_1 and L_2, respectively, is

$$M = k\sqrt{(L_1 L_2)} \qquad (14.2)$$

Example
Determine the coupling coefficient between two coils of inductance 10 mH and 40 mH, respectively, if the mutual inductance is 5 mH.

Solution
From eqn (14.2)

$$k = \frac{M}{\sqrt{(L_1 L_2)}}$$

$$= \frac{5 \times 10^{-3}}{\sqrt{(10 \times 10^{-3} \times 40 \times 10^{-3})}} = 0.25 \text{ (Ans.)}$$

14.3 BASIC PRINCIPLE OF THE TRANSFORMER

The operating principle of the transformer can be described in terms of the diagram in Figure 14.2. We have seen earlier that an e.m.f. is induced in a coil whenever the magnetic flux linking with it changes, the equation for the induced e.m.f. being

$$E = N \frac{\Delta \Phi}{\Delta t}$$

where N is the number of turns on the coil and $\frac{\Delta \Phi}{\Delta t}$ is the rate at which the magnetic flux changes.

However, in the transformer in Figure 14.2, the flux links with both the primary and secondary coils, so that the rate of change of magnetic flux is the same for both coils. That is, there is an induced e.m.f. E_2 in the secondary winding, and a 'back' e.m.f. E_1 in the primary winding. Hence

$$\frac{\Delta \Phi}{\Delta t} = \frac{E_1}{N_1} = \frac{E_2}{N_2} \qquad (14.3)$$

That is, **each winding supports the same number of volts per turn.**

Now, if we can assume that the transformer has no power loss (this is not correct in practice since each winding has some resistance and therefore gives rise to an $I^2 R$ loss [often known as a **copper loss**], and the iron circuit of the transformer has both *eddy-current loss* and *hysteresis loss* [collectively known as the **iron loss**, the **core loss** or as the **no-load loss**]), then we may assume that the power consumed by the load connected to

fig 14.2 *(a) a simple transformer, (b) circuit symbol*

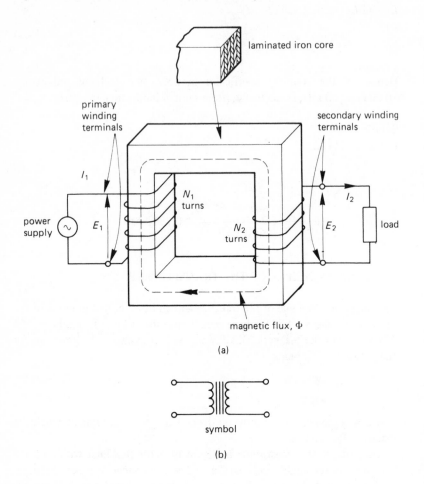

laminated iron core

primary winding terminals

secondary winding terminals

I_1

N_1 turns

I_2

power supply

E_1

N_2 turns

E_2

load

magnetic flux, Φ

(a)

symbol

(b)

the secondary terminals is equal to the power supplied by the primary winding. That is

$$E_1 I_1 \cos \phi = E_2 I_2 \cos \phi$$

where

E_1 = primary winding voltage
E_2 = secondary winding voltage
I_1 = primary winding current
I_2 = secondary winding current
$\cos \phi$ = power factor of the load

The cos ϕ term cancels on both sides of the equation, giving

$$E_1 I_1 = E_2 I_2 \tag{14.4}$$

Eqns (14.3) and (14.4) enable us to spell out two important design considerations, namely

1. **each winding supports the same number of volts per turn** (eqn (14.3));
2. **the volt–ampere product in each winding is the same** (eqn (14.4)).

We can combine the two equations into a composite equation as follows

$$\frac{E_1}{E_2} = \frac{N_1}{N_2} = \frac{I_2}{I_1} \tag{14.5}$$

The ratio $E_2:E_1$ is described as the **voltage transformation ratio** of the transformer (this is the reciprocal of the ratio. $E_1:E_2$ in eqn (14.5)). If this ratio is greater than unity then E_2 is greater than E_1, and the transformer has a **step-up voltage ratio**; if the ratio is less than unity, the transformer has a **step-down voltage transformation ratio**.

The ratio $I_2:I_1$ is the **current transformation ratio**. If $I_2:I_1$ is greater than unity, then I_2 is greater than I_1 and the transformer has a **step-up current ratio**; if $I_2:I_1$ is less than unity, the transformer has a **step-down current ratio**.

A transformer which has a step-down voltage ratio has a step-up current ratio and vice versa. For example, if $E_1 = 400$ V and $E_2 = 20$ V, the transformer has a voltage ratio of $E_2:E_1 = 20:400 = 1:20 = 0.05$ (which is usually described as a *step-down voltage ratio* of 20 to 1 or 20:1). The same transformer has a *step-up current ratio* of 20:1, so that if the current in the primary winding is 0.5 A, the secondary winding current is 10 A.

If we cross-multiply the centre term with the right-hand term in eqn (14.5) we get the equation

$$I_1 N_1 = I_2 N_2$$

which gives another cardinal design rule which is

3. **Ampere–turn balance is maintained between the windings.**

Example

A single-phase transformer supplied from a 1100-V supply has a step-down voltage ratio of $5:1$. If the secondary load is a non-inductive resistor of 10 ohms resistance calculate (a) the secondary voltage, (b) the secondary current and the primary current, (c) the power consumed by the load.

Solution

$$E_1 = 1100 \text{ V}; \frac{E_2}{E_1} = \frac{1}{5}; R = 10 \ \Omega$$

(a) $\dfrac{E_2}{E_1} = \frac{1}{5}$, hence

$$E_2 = \frac{E_1}{5} = \frac{1100}{5} = 220 \text{ V (Ans.)}$$

(b) $\quad I_2 = \dfrac{E_2}{R} = \dfrac{220}{10} = 22 \text{ A (Ans.)}$

Also, since $\dfrac{E_2}{E_1} = \frac{1}{5}$, then $\dfrac{I_2}{I_1} = 5$

therefore

$$I_1 = \frac{I_2}{5} = \frac{22}{5} = 4.4 \text{ A (Ans.)}$$

(c) power consumed by the load $= I_2^2 R = 22^2 \times 10$

$$= 4840 \text{ W (Ans.)}$$

14.4 CONSTRUCTION OF A PRACTICAL TRANSFORMER

The operation of a *two-winding transformer* has been based on the construction in Figure 14.2, in which each winding is on a separate *limb* of the iron circuit. However, this arrangement is not efficient since the two windings are well separated from one another, resulting in a large magnetic leakage with an associated low value of magnetic coupling coefficient.

Efficient power transformer design brings the two windings into intimate contact with one another so that there is a close magnetic coupling between the two. The principal types of magnetic circuit and winding arrangements used in power transformers are described below.

Magnetic circuit design

A practical iron circuit has a certain amount of power loss because of induced eddy current power loss in the iron circuit. Using good engineering design, it is possible to minimise this power loss. The way in which this is done in power transformers is to construct the iron circuit using **thin iron laminations** as shown in Figure 14.2.

The laminations increase the resistance of the iron core to the flow of eddy current, so that the induced eddy-current power-loss in the core is reduced. The laminations are isolated from one another by a very thin coat of varnish (or oxide) on one side of each lamination. At the same time, the total cross-sectional area of the iron circuit is not reduced, so that the magnetic flux density is unchanged.

In the **core-type** magnetic circuit (see Figure 14.3(a)), one half of each winding is associated with each limb of the transformer, the magnetic

fig 14.3 *(a) core-type magnetic circuit construction, (b) shell-type construction*

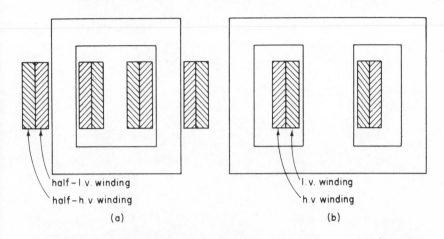

half – l.v. winding
half – h.v winding
(a)

l.v. winding
h.v winding
(b)

circuit having a uniform cross-sectional area throughout. In the **shell-type** magnetic circuit (see Figure 14.3(b)) both the primary and the secondary winding are on the centre limb which has twice the cross-sectional area of the outer limbs.

Winding design

In the **concentric construction** of windings (Figure 14.4(a)), the low-voltage and high-voltage windings are wound concentrically around the iron core. In the **sandwich construction** of windings (Figure 14.4(b)), the high-voltage winding is sandwiched between the two halves of the low voltage winding.

Other general features of transformer construction

In large power transformers, ventilation spaces are left in the windings to allow space for a coolant to circulate, which may be either air or oil. Oil is used as a coolant in large equipment because it is an efficient insulating medium at normal operating temperature.

14.5 AUTOTRANSFORMERS OR SINGLE-WINDING TRANSFORMERS

In some applications, a two-winding transformer of the type described above can be replaced by a simpler single-winding transformer known as an **autotransformer**. Two simplified versions of the autotransformer are shown in Figure 14.5. An autotransformer can either have a step-down voltage ratio (Figure 14.5(a)) or a step-up voltage ratio (Figure 14.5 (b)).

fig 14.4 *(a) concentric winding construction and (b) sandwich winding construction*

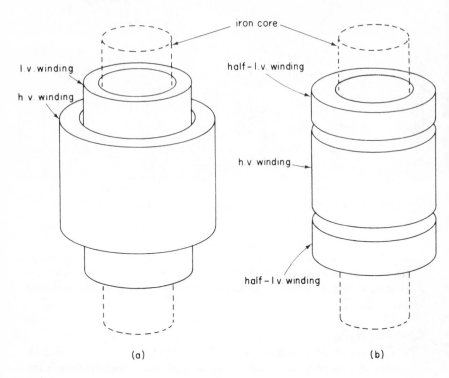

(a)   (b)

fig 14.5 *autotransformer with (a) a step-down voltage ratio, (b) a step-up voltage ratio*

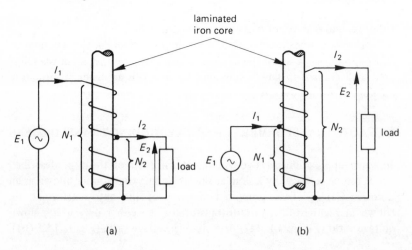

(a)   (b)

The general principles of the transformer apply to the autotransformer, that is

$$\frac{E_1}{N_1} = \frac{E_2}{N_2} \quad \text{and} \quad I_1 N_1 = I_2 N_2$$

For some applications the autotransformer has the drawback that the primary and secondary windings are not electrically isolated from one another (as they are in the two-winding transformer); this introduces the hazard that the low-voltage winding can be accidentally charged to the high voltage potential under certain fault conditions.

14.6 TRANSFORMER EFFICIENCY

The efficiency of a transformer for a particular electrical load is given by the following:

$$\text{efficiency}, \eta = \frac{\text{output power (watts)}}{\text{input power (watts)}}$$

$$= \frac{\text{output power}}{\text{output power} + \text{power losses}}$$

$$= \frac{\text{input power} - \text{losses}}{\text{input power}}$$

A well-designed transformer has relatively low power-losses (low, that is, when compared with the power transmitted to the load), and the efficiency is quite high (95 per cent or better at normal load). The power loss occurring within a transformer can be divided into two types, namely

1. I^2R **loss or 'copper' loss** due to flow of load current in the transformer windings (this power loss varies with the load current).
2. **Power loss in the magnetic circuit** consisting of the hysteresis loss and the eddy-current loss (this loss is fairly constant and is independent of load current, and occurs even under no-load conditions).

Example
A 6600/550 V transformer has an iron loss of 350 W and a full-load copper loss of 415 W. If the full-load secondary current is 45 A at a power factor of 0.6 lagging, calculate the full-load efficiency of the transformer.

Solution
$E_1 = 6600$ V; $E_2 = 550$ V; iron loss = 350 W; full-load copper loss = 415 W; full-load secondary current = 45 A; power factor = 0.6.

Power output from transformer $= E_2 I_2 \cos \phi$

$$= 550 \times 45 \times 0.6 = 14\,850 \text{ W}$$

The power loss in the transformer at full load is given by

copper loss + iron loss = 415 + 350 = 765 W

Hence

full-load efficiency, $\eta = \dfrac{\text{output power}}{(\text{output power} + \text{losses})}$

$$= \dfrac{14\,850}{(14850 + 765)}$$

$$= 0.951 \text{ per unit or } 95.1 \text{ per cent (Ans)}.$$

14.7 INSTRUMENT TRANSFORMERS

In industrial applications, many items of equipment require either a very high voltage to operate them, for example, several kilovolts, or a very high current, for example, hundreds or thousands of amperes. It is impractical in many cases to produce instruments which can be used to measure these values directly and, for this reason, 'instrument' transformers are used to transform the voltage or current level to a practical value which can be measured by conventional instruments (the normal range used is 1 A or 5 A for current and 110 V for voltage).

A **voltage transformer (VT)** or **potential transformer (PT)** is designed to operate with a high resistance voltmeter connected to its secondary terminals. That is, it is designed to operate with a very small load current.

A **current transformer (CT)** is designed to operate with an ammeter connected to its secondary terminals; that is to say, in normal operation the secondary terminals are practically short-circuited (note: since the ammeter has very little resistance, the CT transmits very little power to the ammeter). If the ammeter is disconnected for any reason (such as for maintenance purposes), **the secondary terminals of the CT should be short-circuited**. If this is not done, the primary current can endanger the transformer by causing (i) overheating of the iron core due to the high level of magnetic flux in it, and (ii) damage to the secondary winding by inducing a very high voltage in it.

14.8 USING A TRANSFORMER TO PRODUCE A TWO-PHASE SUPPLY

A two-phase supply is produced at the secondary of the transformer in Figure 14.6. The transformer has a centre-tapped secondary winding and,

fig 14.6 *use of a centre-tapped secondary winding to produce a two-phase (bi-phase) supply*

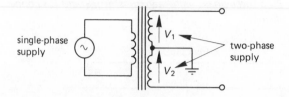

in effect, each half of the secondary winding has the same value of voltage induced in it. However, the voltages V_1 and V_2 are 180° out of phase with one another (see also section 13.2). This type of circuit is used in certain types of rectifier circuit (see Chapter 16), and the same general principle is used to produce a six-phase supply from a three-phase supply (see section 14.10).

14.9 THREE-PHASE TRANSFORMERS

A three-phase transformer can be thought of as a magnetic circuit which has three separate primary windings and three secondary windings on it, either of which may be connected in star or in delta (some three-phase transformers use quite complex connections).

The basis of the three-phase transformer is illustrated in Figure 14.7, in which one phase of a three-phase transformer is shown. The general

fig 14.7 *windings on one limb of a three-phase transformer*

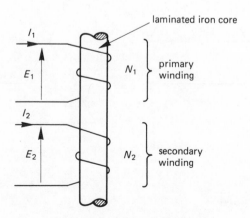

fig 14.8 *(a) a three-phase delta-star transformer and (b) connection diagram*

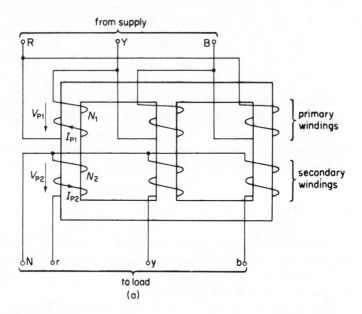

(a)

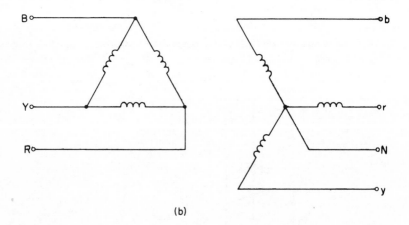

(b)

equation of the single-phase transformer holds good **for each phase of the three-phase transformer**; that is

$$\frac{E_1}{N_1} = \frac{E_2}{N_2}$$

and

$$I_1 N_1 = I_2 N_2$$

the values of E, N and I referred to above are shown in Figure 14.7.

Figure 14.8 shows a three-phase **delta–star** transformer, that is the primary winding is connected in delta and the secondary winding is connected in star. The physical arrangement of the transformer windings are illustrated in Figure 14.8(a) and the circuit diagram in Figure 14.8(b). This type of transformer enables the three-phase, three-wire supply connected to the primary winding to energise a three-phase, four-wire load. This type of transformer is used to connect, say, a local medium-voltage distribution system to a housing estate.

14.10 A THREE-PHASE TO SIX-PHASE TRANSFORMER

Many industrial power electronics circuits need a six-phase power supply. In this section we look at one method of producing a six-phase supply from a three-phase supply.

The transformer described here (see Figure 14.9) uses the method described in section 14.8 for the production of a two-phase supply from a single-phase supply. The basic transformer consists of a delta–star transformer of the type in Figure 14.8(a) with the difference that each secondary winding is centre-tapped. In this way, each of the secondary windings produces two supplies which are 180° out of phase with one another. All the centre-taps of the secondary windings are connected together to form a common neutral point (or star point) for the secondary winding. The net result is a three-phase to six-phase transformer, the secondary 'phase' voltages being displaced 60° from one another.

SELF-TEST QUESTIONS

1. Explain what is meant by the expression 'mutual inductance'. How is mutual inductance utilised in a transformer?
2. What is meant by the 'coupling coefficient' of a magnetic circuit? State the limiting values of the coupling coefficient; explain why its value cannot fall outside these limits.

fig 14.9 *a three-phase to six-phase transformer*

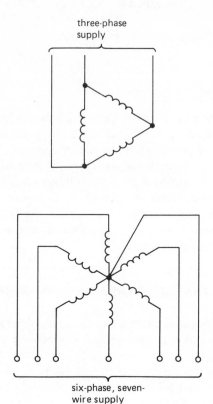

three-phase
supply

six-phase, seven-
wire supply

3. Two coils of equal inductance are magnetically coupled together. If the coupling coefficient is 0.5 and the mutual inductance is 0.5 H, determine the inductance of each coil.

4. A transformer has 50 turns of wire on its primary winding and 10 turns on its secondary winding. If the supply voltage is 100 V, calculate the secondary voltage. A 10-watt load is connected to the secondary terminals; determine the primary and secondary current.

5. Describe the construction and operation of a two-winding transformer. Include in your description details of the magnetic circuit construction and the winding construction.

6. How does an auto-transformer differ from a two-winding transformer?

7. A transformer has a 10:1 step-up voltage ratio. If the efficiency of the transformer is 95 per cent, calculate the power delivered to the load if the primary voltage and current are 100 V and 0.1 A, respectively.

8. Discuss applications of instrument transformers.

SUMMARY OF IMPORTANT FACTS

The **transformer** depends for its operation on the principle of **mutual induction**. The **primary winding** of the transformer is connected to the power source (which must be a.c.), and the load is connected to the **secondary winding**.

The transformer may have either a *single winding* (when it is known as an **autotransformer**) or *more than one winding* (**two-winding** transformers are the most common single-phase transformers). The **iron circuit** of the transformer is **laminated** to reduce the *eddy-current power-loss*. Important rules relating to transformer design are

1. **Each winding supports the same number of volts per turn.**
2. **Ampere–turn balance is maintained between the windings**.

If a transformer has a *step-down voltage ratio* it has a *step-up current ratio*, and vice versa.

The **efficiency** of a transformer is the ratio of the power it delivers to the load to the power absorbed by the primary winding.

Instrument transformers (*potential transformers* or *voltage transformers* and *current transformers*) are used to extend the range of a.c. measuring instruments.

A **three-phase transformer** may be thought of as three single-phase transformers using a common iron circuit. Special winding connections allow a six-phase supply (or a twelve-phase supply for that matter) to be obtained from a three-phase supply.

a.c. MOTORS

15.1 PRINCIPLE OF THE a.c. MOTOR

Imagine that you are looking at the end of the conductor in Figure 15.1(a) when the S-pole of a permanent magnet is suddenly moved from left to right across the conductor. By applying *Fleming's right-hand rule*, you can determine the direction of the induced e.m.f. and current in the conductor. You need to be careful when applying Fleming's rule in this case, because the rule assumes that the *conductor moves relatively to the magnetic flux* (in this case it is the *flux that moves relatively to the conductor*, so the direction of the induced e.m.f. is determined by saying that the flux is stationary and that the conductor effectively *moves to the right*). You will find that the induced current flows away from you.

You now have a current-carrying conductor situated in a magnetic field, as shown in Figure 15.1(b). There is therefore a force acting on the

fig 15.1 *(a) how current is induced in a rotor conductor and (b) the direction of the force on the induced current*

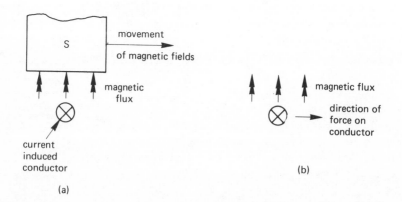

(a)

(b)

conductor, and you can determine the direction of the force by applying Fleming's *left-hand rule*. Application of this rule shows that **there is a force acting on the conductor in the direction of movement of the magnetic field**.

That is, the conductor is accelerated in the direction of the moving magnetic field.

This is the basic principle of the a.c. motor. An a.c. motor therefore provides a means for producing a 'moving' or 'rotating' magnetic field which 'cuts' conductors on the **rotor** or rotating part of the motor. The rotor conductors have a current induced in them by the rotating field, and are subjected to a force which causes the rotor to rotate in the direction of movement of the magnetic field.

15.2 ROTATING AND 'LINEAR' a.c. MOTORS

Most electrical motors have a *cylindrical rotor*, that is, the rotor rotates around the axis of the motor shaft. This type of motor generally runs at high speed and drives its load through a speed-reduction gearbox. Applications of this type of motor include electric clocks, machines in factories, electric traction drives, steel rolling mills, etc.

Another type of motor known as a **linear motor** produces motion in a straight line (known as **rectilinear motion**); in this case the mechanical output from the motor is a linear movement rather than a rotary movement. An application of this type of motor is found in railway trains. If you imagine the train to be 'sitting' above a single metal track (which is equivalent to the 'conductor' in Figure 15.1) and the 'moving magnetic field' is produced by an electromagnetic system in the train then, when the 'magnet' is made to 'move' by electrical means, it causes the system to produce a mechanical force between the electromagnet and the track. Since the track is fixed to the ground, the train is 'pulled' along the conductor.

15.3 PRODUCING A ROTATING OR MOVING MAGNETIC FIELD

A simplified form of a.c. motor is shown in Figure 15.2. We will describe here a very simplified theory of the rotating field that will satisfy our needs. Imagine the three-phase supply to be a sequence of current 'pulses' which are applied to the red, yellow and blue lines in turn. Each current pulse causes one of the coils to produce a magnetic field, and when one electromagnet is excited, the others are not. As the current pulse changes

fig 15.2 *production of a rotating magnetic field*

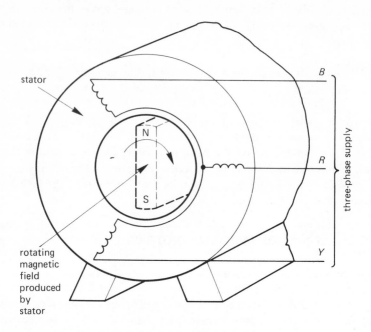

from the red line to the yellow line and then to the blue line, so the magnetic effect 'rotates' within the machine.

The net result is that we can regard this field as the rotating permanent magnet shown dotted in Figure 15.2. The speed at which the magnet rotates is known as the **synchronous speed** of the machine. Various symbols are used to describe the synchronous speed as follows

N_S is the synchronous speed in rev/min

n_S is the synchronous speed in rev/s

ω_S is the synchronous speed in rad/s

The relationship between them is

$$N_S = 60\, n_S$$

and

$$\omega_S = 2\pi n_S = 120\pi N_S \tag{15.1}$$

In practice the synchronous speed of the rotating field is dependent not only on the number of pole-pairs produced by the motor (that is, the number of N- and S-pole combinations – the motor in Figure 15.2 has *one pole-pair*), but also on the frequency, f Hz, of the a.c. supply, the equation relating them is

$$\text{frequency}, f = n_S p \text{ Hz} \tag{15.2}$$

where

f = supply frequency in Hz

n_S = synchronous speed in rev/s

p = number of *pole-pairs* produced by the motor

Example

Calculate the speed of the rotating field of a two-pole motor in rev/min if the supply frequency is (i) 50 Hz, (ii) 60 Hz. What is the synchronous speed if the motor has two pole-pairs, that is, two N-poles and two S-poles?

Solution

(i) $p = 1$; $f = 50$ Hz. From eqn (11.2)

$$n_S = \frac{f}{p} = \frac{50}{1} = 50 \text{ rev/s}$$

and

$$N_S = 60 n_S = 60 \times 50 = 3000 \text{ rev/min (Ans.)}$$

(ii) $p = 1$; $f = 60$ Hz.

$$n_S = \frac{f}{p} = \frac{60}{1} = 60 \text{ rev/s}$$

and

$$N_S = 60 n_S = 60 \times 60 = 3600 \text{ rev/min (Ans.)}$$

Note: Since the US supply frequency is 60 Hz, then a motor with the same number of pole-pairs runs $\frac{60}{50} = 1.2$ times faster than the same motor in the UK.

You can see from eqn (15.2) that if the number of pole-pairs is doubled, then the speed is halved. The respective synchronous speeds for four-pole motors, that is, two pole-pair motors are

$$50 \text{ Hz: } \frac{3000}{2} = 1500 \text{ rev/min (Ans.)}$$

$$60 \text{ Hz: } \frac{3600}{2} = 1800 \text{ rev/min (Ans.)}$$

15.4 A PRACTICAL a.c. MOTOR – THE INDUCTION MOTOR

The simplest form of motor is the **cage rotor induction motor** which has
the rotor structure shown in Figure 15.3. The rotor conductors consist of
a series of stout aluminium or copper **bars** which are short-circuited at the
ends by substantial metal **end rings**. The conductor assembly resembles a
'squirrel cage' and, for this reason, the motor is sometimes described as a
squirrel cage induction motor.

The rotor conductors are embedded in a *laminated iron core* (see Figure
15.3(b) and (c)) which provides the iron path for the magnetic flux produced by the windings on the **stator** or stationary part of the machine; the

fig 15.3 *construction of the rotor of a cage induction motor*

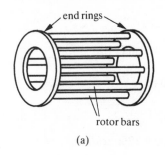

end rings

rotor bars

(a)

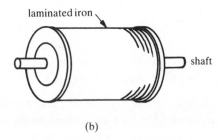

laminated iron

shaft

(b)

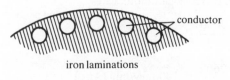

conductor

iron laminations

(c)

rotor is laminated in order to reduce the rotor iron loss. No electrical connections are made to the rotor of the machine.

The stator carries a three-phase winding which produces a rotating magnetic field when it is energised by a three-phase supply.

15.5 'FRACTIONAL SLIP' OF AN INDUCTION MOTOR

When a three-phase supply is connected to the stator winding of the induction motor, a magnetic field is produced which rotates at the synchronous speed of the motor, this flux cutting the rotor conductors. The net result is that a current is induced in the conductors in the manner described in section 15.1, and the rotor experiences a **torque** or *turning moment* which accelerates it in the direction of the rotating field.

Let us examine what happens next. As the rotor speeds up, the *relative speed* between the rotating field (which is the constant synchronous speed) and the rotor reduces. Consequently, the rotating magnetic field does not cut the rotor at the same speed; the net result is that the current is still induced in the rotor, the rotor continues to accelerate.

For the moment, let us assume that the rotor can continue acclerating until its speed is equal to that of the rotating field. In this event, the rotor conductors no longer 'cut' the magnetic field produced by the stator (since they both rotate at the same speed!). Since current is only induced in a conductor when it 'cuts' magnetic flux, it follows in this case that the rotor current falls to zero; that is, the rotor no longer produces any torque and it will begin to slow down. We therefore conclude that, under normal operating conditions, **the rotor of an induction motor will never rotate at the synchronous speed of the rotating magnetic field**.

There are, of course, special conditions under which the rotor rotates at a speed which is equal to or even higher than the synchronous speed, but a discussion of these is outside the scope of this book.

The actual difference in speed (in rev/min, or in rev/s or in rad/s) between the rotor and the rotating field is known as the **slip** of the rotor. It is more usual for engineers to refer to the **fractional slip**, s, which has no dimensions and is calculated as follows:

$$\text{fractional slip, } s = \frac{\text{synchronous speed} - \text{rotor speed}}{\text{synchronous speed}} \qquad (15.3)$$

If the speed is in rev/min, the fractional slip is

$$s = \frac{(N_S - N)}{N_s}$$

If the speed is in rev/s, the fractional slip is

$$s = \frac{(n_S - n)}{n_S}$$

If the speed is in rad/s, the fractional slip is

$$s = \frac{(\omega_S - \omega)}{\omega_S}$$

where N, n and ω is the speed of the rotor in rev/min, rev/s and rad/s, respectively.

Example

Calculate the fractional slip of a three-phase, four-pole machine which has a supply frequency of 50 Hs if the rotor speed is 1440 rev/min.

Solution

$p = 2$ (2 pole-pairs); $f = 50$ Hz; $N = 1440$ rev/min.
From eqn (15.2)

$$n_S = \frac{f}{p} = \frac{50}{2} = 25 \text{ rev/s}$$

hence

$$N_S = 60 n_S = 60 \times 25 = 1500 \text{ rev/min}$$

From eqn (15.3), the fractional slip is

$$s = \frac{(N_S - N)}{N_S} = \frac{(1500 - 1440)}{1500}$$

$$= 0.04 \text{ per unit or 4 per cent (Ans.)}$$

That is to say, the rotor runs at a speed which is 4 per cent below the synchronous speed.

15.6 THE SYNCHRONOUS MOTOR

A number of applications call for a motor whose rotor runs at precisely the same speed as that of the rotating magnetic field. This type of motor is known as a **synchronous motor**.

Quite simply, the principle is to replace the cage rotor with what is equivalent to a permanent magnet. The N-pole of this magnet is attracted to the S-pole of the rotating field, and the S-pole on the permanent magnet is attracted to the N-pole of the rotating field. Provided that the magnetic pull between the permanent magnet and the rotating field is large enough, the rotor runs at the same speed as the rotating field.

In practice, the 'permanent magnet' effect is produced by an electromagnet which is excited by a d.c. power supply which may be either a d.c. generator (known as an **exciter**) or from a rectifier.

Synchronous motors are generally more expensive than induction motors which (apart from providing an absolutely constant speed) have the ability to provide a certain amount of power factor 'correction' to an industrial plant. The latter feature means that the synchronous motor can 'look' like a capacitor so far as the a.c. power supply is concerned. It is this feature which makes the synchronous motor financially attractive to industry. Detailed treatment of this aspect of the operation of the synchronous motor is beyond the scope of this book.

15.7 SINGLE-PHASE INDUCTION MOTORS

As explained earlier, in order to develop a torque in an induction motor, it is necessary to produce a rotating magnetic field. Unfortunately, a coil connected to a single-phase supply can only produce a *pulsating magnetic field*; that is to say, the magnetic field merely pulsates in and out of the coil and does not 'rotate' or sweep past the coil.

A rotating magnetic field can only be produced by using a supply with more than one phase; that is, at least two coils are needed, each of which is excited by a current which has a phase difference with the current in the other coil (in the three-phase case, three coils are used, the currents in the three coils being displaced from one another by 120°). It is therefore necessary to produce a second 'phase' from the single-phase supply in order to start a single-phase motor. One way in which this can be done is described below.

Once a rotating field has been produced, the rotor of the motor begins to rotate in the direction of the rotating field; at this point it is possible to 'switch off' the second phase, and the rotor will continue running so long as one coil is connected to the single-phase supply. This principle is utilised in the **capacitor-start motor** in Figure 15.4.

The single phase supply is connected directly (via the starting switch) to the **running winding**. The 'second phase' is obtained by means of capacitor C; the capacitor has the effect of supplying a current to the **starting winding** which has a phase angle which differs from the current in the running winding. The net result is that the starting and running windings can be thought of as being supplied from a two-phase supply; these produce a rotating magnetic field in the motor (its effect is not as good as a three-phase motor, but is adequate to start the motor).

fig 15.4 *a single-phase, capacitor-start induction motor*

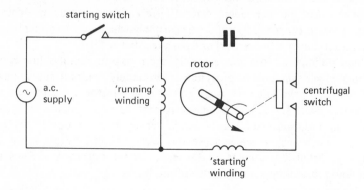

The rotor begins to accelerate and, when it reaches a pre-determined speed set by a centrifugal switch, this switch cuts the starting winding out of circuit. The motor continues to run under the influence of the single-phase supply connected to the running winding.

Single-phase motors are well-suited to small drives used in domestic equipment, but are both uneconomic and inefficient for large industrial drives.

SELF-TEST QUESTIONS

1. Explain how a 'rotating' magnetic field is produced by a polyphase a.c. motor. How can this idea be adopted to produce a 'linear' motor?
2. A 10-pole a.c. motor is connected to a 50-Hz supply. Calculate the synchronous speed of the rotating field.
3. Explain the principle of operation of (i) an induction motor, (ii) a synchronous motor.
4. A 50-Hz, four-pole induction motor has a fractional slip of 5 per cent. Calculate the speed of the motor.
5. With the aid of a circuit diagram, explain the operation of one form of single-phase a.c. motor.

SUMMARY OF IMPORTANT FACTS

a.c. motor action depends on the production of a **rotating magnetic field** (or a *linearly moving magnetic field* in the case of a 'linear' motor).

A **cage rotor induction motor** has a rotor consisting of a cylindrical cage of conductors in laminated iron. The stator (which is also laminated) carries a poly-phase winding which produces a rotating magnetic field when excited by a poly-phase supply. The current induced in the rotor acts on the rotating field to produce a torque which acts **in the same direction as the rotating field**.

The supply frequency, f, is related to the synchronous speed of the rotating field, n_S rev/s, and the number of *pole-pairs*, p, by the equation

$f = n_S p$ Hz

The rotor of an induction motor runs at a speed, n rev/s, which is slightly less than the synchronous speed. The **fractional slip**, s, of an induction motor is given by

$$s = \frac{(n_S - n)}{n_S}$$

The rotor of a **synchronous motor** runs at the same speed as the synchronous speed of the rotating field. The primary advantage of the synchronous motor is that it can draw a current at a leading power factor (the current drawn by an induction motor lags behind the supply voltage), but it has the disadvantage of higher cost and complexity than an induction motor.

Single-phase induction motors are generally more complex than three-phase types because they need both a 'starting' and a 'running' winding, together with some means of making the transition from starting to running conditions.

POWER ELECTRONICS

16.1 SEMICONDUCTORS

A **semiconductor** is a material which not only has a resistivity which is mid-way between that of a good conductor and an insulator, but has properties which make it very useful in the field of electronics and power electronics. There are two principal categories of semiconductor, namely *n*-type semiconductors and *p*-type semiconductors.

An ***n*-type semiconductor** is one whose physical make-up causes it to have **mobile negative charge carriers** in its structure, hence the name *n-type* semiconductor (these negative charge carriers are, of course, *electrons*). When current flows in an *n*-type semiconductor, the current flow is largely due to the movement of negative charge carriers, that is, electrons. Hence, in an *n*-type semiconductor, electrons are known as **majority charge carriers** (since they carry the majority of the current). A small amount of current flow in *n*-type material is due to the movement of **holes** (see Chapter 1 for a discussion on electronic 'holes'); in *n*-type material, holes are described as **minority charge carriers**.

A ***p*-type semiconductor** is one whose physical structure causes it to have **mobile positive charge carriers**, hence the name *p*-type semiconductor (the positive charge carriers are *holes*). The **majority charge carriers** in *p*-type material are holes, and current flow in *p*-type material can be thought of as being largely due to movement of holes. *Electrons* are the *minority charge carriers* in *p*-type material.

A range of semiconductor materials are used by the power electronics industry, the most popular being silicon, closely followed by germanium. A range of specialised items of equipment are manufactured using such materials as cadmium sulphide, indium antimonide, gallium arsenide, bismuth telluride, etc.

16.2 SEMICONDUCTOR DIODES (THE *p–n* JUNCTION DIODE)

A **diode** is an electronic device which allows current to flow through it with little opposition for one direction of flow, but prevents current flow in the reverse direction. That is, it is the electrical equivalent of a 'one-way' valve.

It has *two electrodes* (see Figure 16.1), known respectively as the **anode** and the **cathode**; the anode is a *p*-type semiconductor and the cathode is *n*-type (a diode is a *single piece of semiconductor* with an *n*-region and a *p*-region; it *is not* simply a piece of *p*-type material which has been brought into contact with a piece of *n*-type material). The symbol representing a diode is shown in the lower part of Figure 16.1; the 'arrow' shape which represents the anode *points in the 'easy' direction of current flow*. That is

the diode offers little resistance to current flow when the anode is positive with respect to the cathode; the diode offers nearly infinite resistance to current flow when the anode is negative with respect to the cathode.

You may therefore regard a diode as a 'voltage-sensitive' switch which is switched to the 'on' or is closed when the anode is positive with respect to the cathode, and is 'off' or open when the anode is negative.

fig 16.1 *a p–n junction diode*

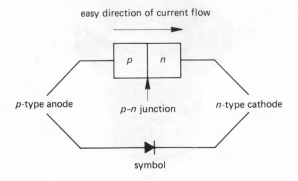

The diode in Figure 16.1 is known as a *p–n* **junction diode**, and a practical diode of this type is physically very small (even a 'power' diode which can handle several hundred amperes is only a few millimetres in diameter!). However, each diode is housed in a container or canister of manageable size for two reasons:

1. it must be large enough to handle;
2. it must have sufficient surface area to dissipate any heat generated in it.

A photograph of a 'signal diode' (a small-current diode) is shown in Figure 16.2(a), its diameter being 2–3 mm. A 'power diode' is shown in Figure 16.2(b); this is seen to be physically larger than the signal diode, and is suitable for bolting onto a **heat sink** which is used to dissipate the heat produced by the diode in normal use.

fig 16.2   *photograph of (a) a signal diode and (b) a power diode*

(a)

(b)

Reproduced by kind permission of RS Components

A heat sink is a metal (often aluminium) finned structure to which the diode is fastened. It may be painted black to increase its heat radiation capacity and, in some cases, may be fan-coiled. A selection of heat sinks are shown in Figure 16.3.

fig 16.3 *a selection of heat sinks*

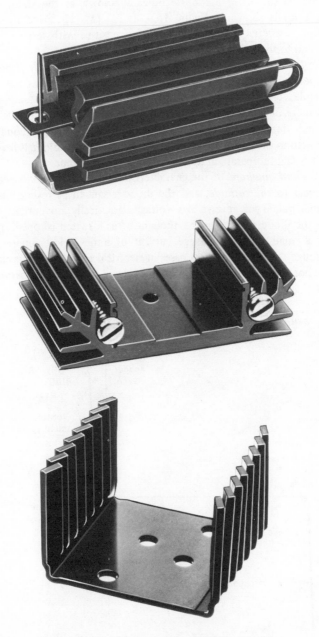

16.3 DIODE CHARACTERISTIC CURVES

The characteristic of a diode (technically known as the **static anode characteristic**) is shown in Figure 16.4. It has two operating areas corresponding respectively to the anode being positive with respect to the cathode and to the anode being negative.

In the *first quadrant* of the graph in figure 16.4, the *anode is positive with respect to the cathode*, and this is known as **forward bias** operation. In this mode, current flows easily through the diode, and the p.d. across the diode is more or less constant between about 0.3 V and 0.5 V (the value of this p.d. depends to a great extent on the type of semiconductor material used in the construction of the diode). When the diode is forward-biassed it is said to operate in its **forward conduction mode**.

In the *Third quadrant* of the characteristic curve, the *anode is negative with respect to the cathode*, and the diode is said to be **reverse biassed**. At normal values of reverse-bias voltage practically no current flows through the diode and it can be thought of as a switch is 'open' or off. In fact, a small value of **leakage current** of a few microamperes does flow through the diode; in most cases in electrical engineering this current can be ignored (this current is sometimes known as the **reverse saturation current**).

fig 16.4 *static anode characteristic of a diode*

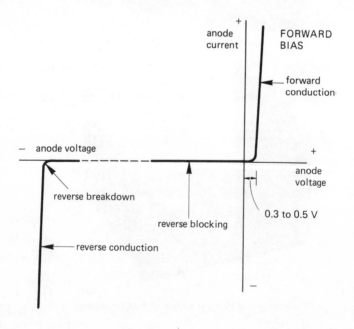

If the reverse bias voltage is increased to a sufficiently high value, **reverse conduction** commences at a voltage known as the **reverse breakdown voltage**. Any attempt to increase the reverse bias voltage further results in a rapid increase in the current through the diode. If the current through the diode is not restricted in value (by, say, a fuse), the diode will rapidly overheat and will fail catastrophically.

If must be pointed out that diodes used in power supplies are operated well below their rated reverse breakdown voltage, so that the possibility of failure of the above reason is unlikely.

A branch of the diode family known as **Zener diode** is operated in the reverse breakdown mode, and details are given in *Mastering Electronics* by J. Watson.

16.4 RECTIFIER CIRCUITS

A **rectifier** is a circuit containing diodes which *convert alternating current into direct current (unidirectional current)*. The basic arrangement is shown in Figure 16.5; alternating current enters the rectifier from the a.c. supply and leaves as d.c.

The a.c. supply may be either a single-phase supply, or a three phase supply, etc. The rectifier circuit can be any one of a number of possible types of circuit including half-wave or full-wave; these terms are described in the following sections

fig 16.5 *block diagram of a rectifier*

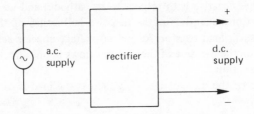

16.5 SINGLE-PHASE, HALF-WAVE RECTIFIER CIRCUIT

A **half-wave rectifier circuit** is one which utilises one-half of each complete wave or cycle of the a.c. supply in order to send current to the 'd.c.' load. A typical single-phase, half-wave rectifier circuit is shown in Figure 16.6(a).

The operation of the circuit is described below, and the corresponding waveform diagrams are shown in diagram (b), (c), and (d) of Figure 16.6.

During each *positive half-cycle* of the a.c. wave (that is, in the time intervals *A–B* and *C–D* of Figure 16.6(b)), the diode is *forward biassed*

fig 16.6 *single-phase, half-wave rectifier circuit*

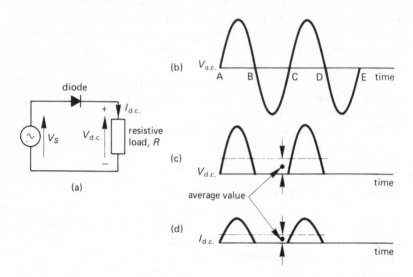

(that is, its anode is positive with respect to its cathode) and current flows through it. Since the 'd.c.' load is resistive, the current in the load is proportional to the voltage (Ohm's law), and the voltage across the load is the same as that of the alternating wave, that is, the 'd.c.' voltage is sinusoidal for the first half-cycle (see Figure 16.6(c)).

During each *negative half-cycle* of the a.c. wave (that is, during the time intervals *B–C* and *D–E* in Figure 16.6(b)), the diode is *reverse biassed* (that is, its anode is negative with respect to its cathode) and no current flows through it. Consequently, in the negative half-cycles of the supply, no current flows in load resistor, *R*, and no voltage appears across it. Hence the current on the d.c. side of the rectifier flows in one direction only, that is, it is **unidirectional**.

The value of the d.c. voltage, $V_{d.c.}$, across the load is calculated from the equation

$$V_{d.c.} = \frac{V_m}{\pi} = 0.318V_m = 0.45V_s \qquad (16.1)$$

where V_m is the maximum value of the a.c. supply voltage, and V_s is the r.m.s. value of the a.c. voltage. The d.c. current in the load is calculated from

$$I_{d.c.} = \frac{V_{d.c.}}{R} \qquad (16.2)$$

where *R* is the resistance of the load in ohms.

Example

A 240-V sinusoidal supply is connected to a single-phase, half-wave rectifier circuit of the type in Figure 16.6(a). Calculate (i) the direct voltage across the load and (ii) the current in a load resistance of 100 ohms. Determine also (iii) the power consumed by the load.

Solution

V_s = 240 V (r.m.s.); R = 100 ohms.

(i) $V_{d.c.}$ = $0.45V_s$ = 0.45 × 240 = 108 V (Ans.)

(ii) $I_{d.c.}$ = $\dfrac{V_{d.c.}}{R}$ = $\dfrac{108}{100}$ = 1.08 A (Ans.)

(iii) d.c. power = 1.08^2 × 100 = 116.6 W (Ans.)

16.6 SINGLE-PHASE, FULL-WAVE RECTIFIER CIRCUITS

A **full-wave rectifier circuit** utilises both half-cycles of the a.c. supply wave, so that current flows in the d.c. side of the circuit during both half-cycles of the a.c. wave. With the same supply voltage and load resistance, this has the effect of doubling the d.c. voltage when compared with the half-waves case and that the d.c. current is doubled.

The popular forms of single-phase, full-wave rectifier circuit are the *centre-tap (or bi-phase)* circuit and the *bridge* circuit, and are described in the following paragraphs.

Single-phase centre-tap (or bi-phase) rectifier circuit)

The circuit is shown in Figure 16.7(a). It contains two diodes $D1$ and $D2$ which are energised by a transformer with a **centre-tapped secondary winding**. You will recall that it was shown in Chapter 14 that this type of winding provides a two-phase output voltage, the voltage at points X and Y being of opposite phase to one another (centre point Z being the 'common' zero volts line).

In one half-cycle of the supply, the voltage at X is positive with respect to Z on the secondary winding, and Y is negative. This causes D1 to be forward biassed at this time and D2 to be reverse biassed; the net result is that D1 conducts and D2 is cut off. Current therefore flows out of the positive terminal and into the load. In the other half-cycle of the supply, the polarity at X and Y are reversed so that D1 is reverse biassed and D2 is forward biassed. During these half-cycles, current flows through D2 and

fig 16.7 *single-phase, full-wave (a) centre-tap circuit and (b) brige circuit*

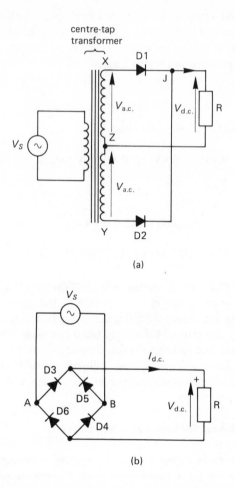

(a)

(b)

into the load via the positive terminal. That is, current always flows through the load in the same direction during both half-cycles of the a.c. waveform.

The corresponding voltage and current waveforms are shown in Figure 16.8. The current through the diodes is 'summed' at junction J in Figure 16.7 to give a pulsating but unidirectional (d.c.) load current.

The d.c. (average) load voltage, $V_{d.c.}$, is calculated from the equation

$$V_{d.c.} = 2V_m/\pi = 0.636V_m = 0.9V_{a.c.} \qquad (16.3)$$

fig 16.8 *voltage and current waveforms for a single-phase, full-wave rectifier circuit*

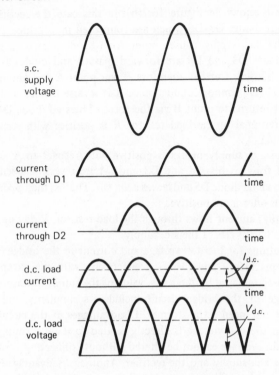

where V_m is the maximum value of the a.c. voltage induced in one-half
of the secondary winding (see Figure 16.7(a)), and $V_{a.c.}$ as its r.m.s.
value. The d.c. load current, $I_{d.c.}$ is given by

$$I_{d.c.} = V_{d.c.}/R \qquad (16.4)$$

where R is the resistance of the load in ohms.

A feature of this circuit is that you can select a direct voltage of your
choice simply by obtaining a transformer with the correct voltage transformation ration. Another feature of the circuit is that the transformer enables
the secondary circuit to be electrically 'isolated' from the mains supply;
this feature is particularly useful in special cases such as medical installations.

A possible disadvantage of this circuit over a number of other rectifier
circuits is that it needs a bulky and relatively expensive transformer. This
disadvantage usually means that the centre-tap circuit is only used where
other lighter and less expensive circuits cannot be used.

Single-phase bridge rectifier circuit

The circuit is shown in Figure 16.7(b); in this case the rectifier circuit contains four diodes D3–D6 which are connected in a 'bridge' formation.

When the a.c. supply causes point A to be positive with respect to point B, diodes D3 and D4 are forward biassed and diodes D5 and D6 are reverse biassed. Current therefore leaves point A and enters the load via diode D3 (D6 cannot conduct since it is reverse biassed at this time). The current returns to point B via the forward-biassed diode D4. That is, the upper terminal of the load resistor R is positive with respect to its lower terminal.

When the a.c. supply makes B positive with respect to A, diodes D5 and D6 are forward biassed and D3 and D4 are reverse biassed. Current enters the load via diode D5 and leaves it via D6. That is, the upper terminal of the load is once more positive.

In this way, current flows through the load resistor, R, *in one direction only* in both half-cycles of the a.c. supply.

The equations for the d.c. voltage and current in the bridge circuit are given by eqns (16.3) and (16.4), respectively, with the exception that $V_{a.c.}$ should be replaced by the r.m.s. value of the supply voltage, V_s.

Advantage of the bridge circuit include its simplicity and the fact that it does not need a transformer. **Disadvantages of the circuit** include the fact that the d.c. voltage is directly related to the a.c. supply voltage, that is the d.c. voltage cannot be altered without connecting a transformer between the a.c. supply and the rectifier. Another disadvantage is that the a.c. and d.c. voltages are not 'isolated' from one another; that is to say, if one of the a.c. supply lines is earthed, then you cannot earth one of the d.c. lines.

16.7 SMOOTHING CIRCUITS

The 'd.c.' voltage waveform from simple rectifier circuits of the type described in section 16.6 is in the form of 'pulses.; whilst this may be satisfactory for a number of applications, it is unsuited to applications which need a 'smooth' d.c. voltage. A number of simple low-cost **smoothing circuits** or **ripple filter circuits** using combinations of L and C are widely used to reduce the current and voltage ripple in the d.c. output of rectifier circuits.

Since the single-phase, half-wave rectifier conducts for only one half of each cycle, *it is not a practical proposition to try to obtain a 'smooth' output from it*. We therefore concentrate on circuits primarily used to smooth the output from a full-wave rectifier circuit.

A simple **choke-input filter** is shown in Figure 16.9(a) and you will no doubt recall (see section 11.9 for details) that the function of the inductor or 'choke' is to introduce a high impedance to the 'alternating' ripple current component, and that the function of the reservoir capacitor is to short-circuit the tipple current from the d.c., load resistor. The net result is a significant reduction in the ripple component of the current in the d.c. load (see Figure 16.9(c)).

An improved filter circuit known as a π-**filter** is shown in Figure 16.9(d). This circuit uses two capacitors which are often equal in value and, for a given application, each has about twice the capacitance of C in Figure 16.9(a) (for a given application, the same value of L can be used in both circuits).

16.8 THE THYRISTOR

A **thyristor** is a *multi-layer semiconductor device*. There are two broad categories of thyristor, namely the *reverse blocking thyristor* and the *bidirectional thyristor* or *triac* (the latter being a trade name). Each type is described in this section.

fig 16.9 *(a) choke-input filter circuit and (b) a π-filter*

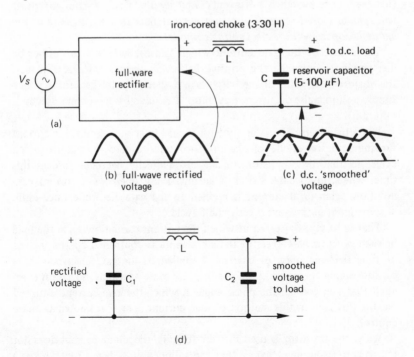

The reverse blocking thyristor

This is the type of device to which engineers generally refer when they mention the 'thyristor'; it is a four-layer semiconductor device which has the construction shown in Figure 16.10(a).

It has three electrodes, namely an **anode**, a **cathode** and a **gate**. The anode and cathode can be thought of in the same way as the anode and cathode of a diode; that is to say, the 'easy' direction of current flow is from the anode to the cathode. It is the four-layer construction and the 'gate' electrode which account for the difference between the thyristor and the diode.

The operation of the thyristor can be explained in terms of the simplified equivalent circuit in Figure 16.11. The thyristor can be thought of as a diode in series with an electronic switch, S, which is controlled by the signal applied to the gate electrode as follows:

> **When the gate current is zero, switch S is open and no current can flow through the thyristor. When gate current flows, switch S closes and current can flow through the thyristor (but only from the anode to the cathode).**

However, once switch S has been closed by the flow of gate current, *it will remain closed (even if the gate current falls to zero) so long as the anode is positive with respect to the cathode.*

The source of the gate current which 'closes' switch S can either be d.c. or a.c. (it must be the positive half-cycle of an a.c. wave!). In fact, the majority of gate 'driver' circuits are special pulse generator which supply a short pulse of current (say about 1 A or so for a few microseconds duration).

When the reverse blocking thyristor is used in an a.c. circuit, the current flowing into the anode of the thyristor falls to zero once during each cycle, so that the thyristor current automatically becomes zero at this time. When this occurs, switch S automatically opens, and current cannot flow again until current is applied to the gate electrode once more at some point in the next positive half-cycle.

That is to say, the point at which the current starts flowing in the load in each *positive half cycle* of the a.c. supply is controlled by the instant of time that the pulse of current is applied to the gate electrode. It is by controlling the point in the positive half cycle at which the gate is energised that you control the phase angle at which the load current is turned on; for this reason, this method of load current control is known as **phase control**.

When the thyristor is used in a d.c. circuit, the anode current does not fall to zero (assuming, that is, that the supply voltage is not switched off)

fig 16.10 *(a) sectional diagram of a p-n-p-n reverse blocking thyristor, (b) circuit symbols and (c) a practical thyristor*

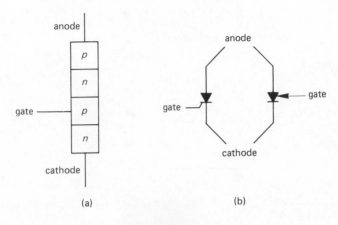

(a) (b)

(c)

Published by kind permission of RS Components

fig 16.11 *simplified operation of the thyristor*

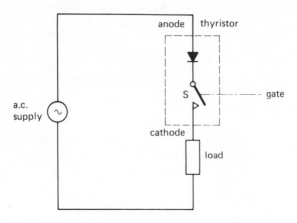

and special current 'commutating' circuits have to be used in order to turn the thyristor off.

The characteristic of a reverse blocking thyristor is shown in Figure 16.12. In the **forward-biassed mode** (when the anode is positive with respect to the cathode) the thyristor has one of two operating conditions, namely

fig 16.12 *characteristic of a reverse blocking thyristor*

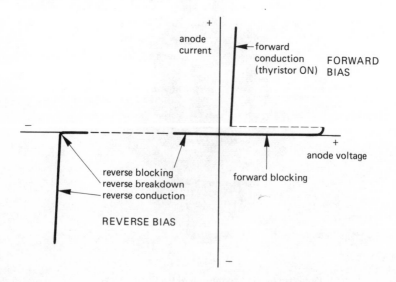

1. when the gate current is zero: in this case the thyristor 'blocks' the flow of current, that is, switch *S* in Figure 16.11 is open. This is known as the **forward blocking mode**;

2. when the thyristor is triggered: in this case the gate current is either flowing or it has just stopped flowing. This causes switch S in Figure 16.11 to close, allowing current to flow through the thyristor. This is known as the **forward conducting mode**.

In the **reverse biassed mode** (when the anode is negative with respect to the cathode) the thyristor blocks the flow of current (and is also known as the **reverse blocking mode**). At a high value of reverse voltage which is well in excess of the voltage rating of the thyristor, reverse breakdown occurs; the thyristor is usually catastrophically damaged if this happens.

A typical application of a reverse blocking thyristor would be to an industrial speed control system, such as a steel rolling mill or an electric train. The operators control lever is connected to a potentiometer which is in a pulse generator in the gate circuit of the thyristor. Altering the position of the control lever has the effect of altering the 'pulse angle' at which the pulses are produced; in turn, this has the effect of controlling the current in (and therefore the speed of) the motor being controlled.

On the whole, the reverse blocking thyristor is more 'robust' than the triac (see next paragraph) and the thyristor can be used in all electric drives up to the largest that are manufactured.

The triac or bidirectional thyristor

This device is one that can conduct in two 'directions' and, although it has a more complex physical structure and operating mechanism than the reverse blocking thyristor, it still has three electrodes. These are known respectively as T1, T2 and the **gate**; circuit symbols for the triac are shown in Figure 16.13.

fig 16.13 *circuit symbols for a triac or bidirectional thyristor*

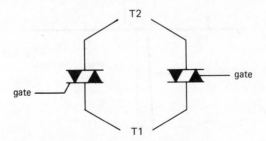

Since the triac can conduct in either direction, terminal T2 may be either positive or negative with respect to terminal T1 when current flow takes place. However, the triac must be triggered in to conduction by the application of a pulse to the gate electrode; the trigger pulse in this case may have either a positive or negative potential with respect to the 'common' electrode T1.

Applications of the triac are limited to power levels up to about a few hundred kilowatts, but this is a situation which is continually changing as technology advances. A popular application of the triac is to a domestic lighting control; the triac and its gate pulse circuitry is small emough to be housed in a standard plaster-depth switch, the knob on the front of the switch controlling the gate pulse circuitry. The knob is connected to a potentiometer which applies 'phase control' to the pulse generator. At switch-on, the triac gate pulses are 'phased back' to between about 150° and 170° so that current only flows for the final 30° to 10° of **each half-cycle** of the supply (remember, the triac conducts for both polarities of T2); this results in the lamp being dimly illuminated at switch-on. As the control knob is turned, the gate pulses are gradually 'phased forward' so that the triac fires at an earlier point in each half-cycle of the supply waveform, resulting in increased illumination. Finally, when the gate pulses are phased forward to 0°, the triac conducts continuously and the lamp reaches its full brilliance.

The characteristic of a triac is shown in Figure 16.14. Before gate current is applied (for either polarity of voltage between T2 and T1) the triac blocks the flow of current (shown as **forward blocking** and

fig 16.14 *characteristic of a triac*

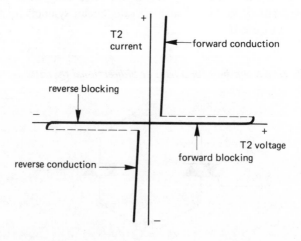

reverse blocking in Figure 16.14). After the application of a gate pulse, the triac conducts for either polarity of potential across it (shown as forward conduction and reverse conduction in Figure 16.14); the triac continues conducting (even when the gate pulse is removed) so long as a p.d. is maintained across it. When the supply current falls to zero (as it does in an a.c. system), the triac reverts to one of its blocking modes until it is triggered again.

16.9 A 'CONTROLLED' THREE-PHASE POWER RECTIFIER

A three-phase 'controlled' bridge rectifier circuit is shown in Figure 16.15, and consists of three thyristors and three diodes, the pulses applied to the gates of the thyristors are 'separated' by the equivalent of 120°, so that only one thyristor conducts at a time.

When TH1 is triggered, current passes through it to the positive terminal of the load, the returns to the supply via either D2 or D3. In fact, only one of the diodes conducts at any one time, which one it is depends on which of the Y or B supply lines is at the most negative potential. That is, current will return via D2 for one period of time, and via D3 for another period of time.

When TH2 is triggered into operation, current returns either via D1 or D3; when TH3 conducts, current returns either via D1 or D2.

Once again, the output current and voltage are controlled by *phase control*, that is the phase angle of the gate pulses is altered via the gate pulse generator.

The circuit in Figure 16.13 is known as a **half-controlled rectifier** because one-half of the devices in the circuit are thyristors.

fig **16.15** *a controlled three-phase bridge rectifier*

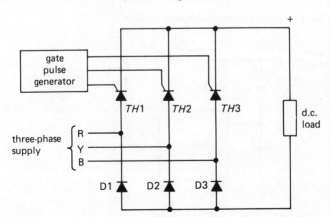

16.10 INVERTORS

An **invertor** is a circuit which **converts d.c. into a.c.**; for example, the circuit which provides the power from the battery of a bus to its fluorescent lights is an invertor. This circuit takes its d.c. supply from a 12-V battery and converts it into a higher voltage a.c. supply for the fluorescent lights.

For a circuit to be able to 'invert'. all the semiconductor elements in the invertor must be thyristors. For example, if all six devices in the half-controlled 'rectifier' in Figure 16.15 were thyristors, then it could act as a three-phase invertor. In this case the 'd.c. load' would be replaced by a battery or d.c. generator, and the 'three-phase supply' would be replaced by a three-phase load such as a motor or heating element.

16.11 A STANDBY POWER SUPPLY

A number of installations need a power supply which is 100 per cent reliable. One method of providing such a power supply is shown in Figure 16.16.

Under normal operating conditions, the load is supplied directly from the mains via contact A of the electronic switch S (which would probably by a thyristor circuit). Whilst the main circuit is working normally, diode D trickle-charges the 'standby' battery.

When the power supply fails, the contact of switch S changes to position B, connecting the load to the output of the invertor circuit. Since the invertor is energised by the standby battery, the a.c. power supply to the load is maintained at all times.

fig **16.16** *one form of standby power supply*

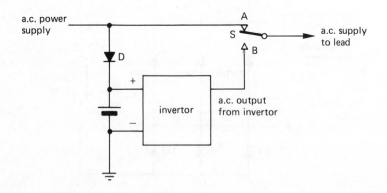

SELF-TEST QUESTIONS

1. What is meant by a p-type semiconductor and an n-type semiconductor? Also explain the meaning of the expressions 'majority charge carrier' and 'minority charge carrier'.
2. Draw and explain the characteristic of a p-n junction diode. What is meant by 'forward bias' and 'reverse bias' in connection with a p-n diode?
3. In what respect do 'half-wave' and 'full-wave' rectifiers differ from one another?
4. A rectifier circuit is supplied from a 100-V r.m.s. supply. For (i) a half-wave and (ii) a full-wave rectifier circuit, calculate the no-load d.c. output voltage. Determine also the load current in each case for a 100-ohm load.
5. Explain the purpose of a 'smoothing' circuit or 'ripple' filter. Draw a circuit diagram for each of two such circuits and explain how they work.
6. What is a thyristor? Draw and explain the shape of the characteristic for (i) a reverse blocking thyristor and (ii) a bidirectional thyristor. Discuss applications of the two types of device.
7. Explain what is meant by an 'invertor'. Where might an invertor be used?

SUMMARY OF IMPORTANT FACTS

A **semiconductor** is a material whose resistivity is mid-way between that of a conductor and that of an insulator; popular semiconductor materials include silicon and germanium. The two main types of semiconductor are **n-type** and **p-type;** n-type has **mobile electrons** in its structure whilst p-type has **mobile holes**. In an n-type material, *electrons* are the **majority charge carriers**, and p-type *holes* are the majority charge carriers.

A **diode** permits current to flow without much resistance when the p-type anode is positive with respect to the n-type **cathode**. In this mode it is said to be **forward biassed**. The diode is **reverse biassed** when the anode is negative with respect to the cathode; in this mode the diode **blocks** the flow of current through it. **Reverse breakdown** occurs if the reverse bias voltage exceeds the reverse breakdown voltage of the diode; the diode can be damaged if the current is not limited in value when reverse breakdown occurs. Diodes designed to work in the reverse breakdown mode are known as **Zener diodes**.

A **rectifier circuit** converts an a.c. supply into d.c. The circuit may either be *single-phase* or *poly-phase*, and may either be *half-wave* or *full-*

wave. The 'ripple' in the output voltage or current from a rectifier can be reduced by means of a **smoothing circuit** or a **ripple filter**.

An **invertor** is the opposite of a rectifier, and converts d.c. into a.c.

A **thyristor** is a multi-layer semiconductor device. A **reverse-blocking thyristor** (which is the type referred to when thyristos are discussed) allows you to control the flow of current from the anode to the cathode by means of a signal applied to its *gate electrode*. A **bidirectional thyristor** (often called a **triac**) allows you to control the flow of current through it in either direction by means of a signal applied to its gate electrode. The gate signal of both types of thyristor may be either d.c. or a.c., or it may be a pulse.

SOLUTIONS TO
NUMERICAL PROBLEMS

Chapter 1   4. 0.2 A
5. 100 S
6. non-linear
7. 1 kHz ; $10\,\mu$ s
8. 400 C; 400 W; 8000 J

Chapter 2   5. 0.1 Ω

Chapter 3   3. 5.51 Ω
4. 0.181 S
5. 86.4 Ω
6. (i) 40 Ω, (ii) 4 Ω
(iii) series = 400 V, parallel = 40 V
(iv) series = 4kW, parallel = 400 W
7. I_1 = 0.136A; I_2 = 0.318A; $I + 1_2$ = 0.454A

Chapter 4   6. 33.33 cm from one end and 66.67 cm from the other end.

Chapter 5   1. 144 MJ; 40 kWh

Chapter 6   2. 1 kV
4. (i) 10 V, (ii) 100 kV
6. 3.54 nF
7. (i) 12 μF ; (ii) 1.091 μF
8. (i) 120 μC ; 10.91 μC ; (ii) 600 μJ; 54.55 μJ
9. (i) 0.01 s ; (ii) 0.01 A ; (iii) 0.007 s ; (iv) 0.05 s ;
(v) 10 V , 0.0 A

Chapter 7   2. 25 000 At; 53 052 At/m ; 0.209 mWb ; 0.209 T
5. 1293 At/m ; 2328 At
8. 125 J
9. (i) 0.5 s ; (ii) zero; 1.0 A ; (iii) 0.35 s (iv) 2.5 s ;
(v) 2.5 J

Chapter 8   1. 200 Hz
2. (i) 1.047 rad ; (ii) 2.094 rad ; (iii) 91.67°
(iv) 263.6°

Chapter 9   3. 0.372 T
4. 960 Nm

Chapter 10   1. 6.58 ms ; 4 MHz
2. (i) 11.79 A ; (ii) 0.174 A, 10.21 A , −4.03 A,
−2.05 A, 9.92 A, 10.72 A , −8.92 A
3. 76.44 V ; 84.84 V
5. (i) 323.9 V leading the 150-V wave by 25.88° ;
(ii) either 141.7 V lagging the 150-V wave by 86.5°
(if the 200-V wave is subtracted), or 141.7 V
leading the 200-V wave by 93.48° if the 150-V
wave is subtracted)
6. (i) 12 500 VA ; (ii) 36.87°, 0.8 ; (iii) 7500 VAr

Chapter 11   1. 0.833 A ; 288Ω
2. Circuit A : (i) 0.667 A ; (ii) I lags behind V_s by 90° ;
Circuit B : (i) 1.25 A ; (ii) I leads V_s by 90°
3. (i) 157.1 Ω, 0.636 A ; (ii) 188.5 Ω , 0.531 A ;
(iii) 314.2 Ω, 0.0318 A
4. 10 μF
5. (i) 14.71 Ω ; (ii) 680 Ω

Chapter 12   1. 628.3 Ω ; 79.6 Ω ; 557.7 Ω
2. 17.93 mA ; V_R = 1.793 V , V_L = 11.27 V ,
V_C = 1.43 V ; P = 32.15 mW, power factor = 0.179
(I lagging V_s)
3. (i) I_R = 0.1 A , I_L = 15.9 mA, I_C = 0.126 A ;
(ii) 0.149 A ; (iii) 47.75° (I leading V_s), 0.672 ;
(iv) 1.49 VA, 1W, 1.105 VAr
5. 0.507 μF; 1.5 A
6. 31.42

Chapter 13   4. 19.05 kV, 173.2V
6. (i) 21.87 A, 21.87 A ; (ii) 21.87 A, 12.63 A
7. 36.87° ; 125 kVA ; 75 VAr

Chapter 14   3. 1 H
4. 20 V ; 0.5 A ; 0.1 A
7. 9.5 W

Chapter 15   2. 5 rev/s or 300 rev/min
4. 1425 rev/min

Chapter 16   4. (i) 45 V, 0.45 A ; (ii) 90 V, 0.9 A

INDEX